Screwpull®

Creation & History of a High-tech Corkscrew

Donald Minzenmayer

Schiffer Publishing Ltd®

4880 Lower Valley Road • Atglen, PA 19310

Dedication

Brother Anthony George Diener (1910–2004) was known to his many friends as Brother Tim. Brother Tim was a priest in the community of the Christian Brothers, a community within the Catholic Church. Brother Tim was overseer of the vineyards and was cellarmaster/winemaker of the Christian Brothers Winery in Napa Valley, California. Brother Tim lived and worked in the Napa Valley for more than 70 years. In his words he was a priest of the land. He was a true expert on how to grow fine grapes and make great wines. He amassed a world-class corkscrew collection and was a founding member of the International Correspondence of Corkscrew Addicts, its first Right and appointed Chaplain. A prayer by Brother Tim preceeded every dinner and lunch at the Annual General Meetings. We all looked forward to his serious and imaginative prayers with thoughts for the occasion.

Brother Tim was a dear friend and an inspiration to all who knew him. In 1979, I sent him one of the new Screwpulls, and his response was: "Wow, I have never removed a cork with such ease or speed. What a great corkscrew."

His advice, whether it was on growing grapes, making wine, studying corkscrews or living life, was highly respected. He was an exceptional man who we all miss.

Brother Timothy Diener with part of his corkscrew collection, c. 1984.

Contents

Foreword....5

Acknowledgments....6

Introduction to Screwpull, Houston, Texas, and Rice University....7

Chapter One—Herbert A. Allen - The Inventor....9

Chapter Two—Screwpull Prototypes....19

Chapter Three—The Screwpull Production Models....37

Chapter Four—Screwpull Knockoffs....49

Chapter Five—The Leverpull Prototypes....63

Chapter Six—Leverpull Production Models 1981–1991 The Hallen Company...79

Chapter Seven—Screwpull Champagne Star....91

Chapter Eight—The Cork Ejector....97

Chapter Nine—The Screwpull Foil Cutter....99

Chapter Ten—The Champagne Seal....103

Chapter Eleven—Patent Drawings....107

Chapter Twelve—Packaging and Marketing....129

Chapter Thirteen—Screwpull 1991–2001 Le Creuset....137

Chapter Fourteen—The Patent Expires....154

Chapter Fifteen—Leverpull Knockoffs....155

Chapter Sixteen—Other Related Items....169

The Last Word....176

MARCUS JAMES
BLACK SWAN Vineyards
SUTTER HOME
BLACK SWAN
MIS EN
CRANELAKE
STE. GENEVIEVE
MARCUS JAMES
BOHEMIAN highway
WINKING OWL Vineyards
CRANELAKE
SUTTER HOME
CRANE LAKE
WINKING OWL
CONCHA Y TORO
DES VOIGNE CELLARS
CRANELAKE
WINKING OWL Vineyards
RODNEY STRONG VINEYARDS
MARCUS JAMES

Foreword

Cork has been used for more than two hundred years to seal wine bottles or jugs. No other material has this great ability to remain elastic over long periods of time, be tasteless and remain odorless. The tricky part is to remove the cork from the bottle or remove the bottle from the cork. The invention of the bottlescrue or corkscrew around the middle of the eighteenth century brought on the development of many designs and variations for removing corks from bottles for the next two centuries. These designs ranged from the very simple straight pull, brute force T-handle corkscrew to very complex mechanical advantage devices. However, fundamentally the corkscrew has not changed in all this time. Screw a screw into the cork and somehow pull it out.

In 1979, the Screwpull was launched in Houston, Texas, invented by a Professional Engineer named Herbert A. Allen. His work on the Screwpull began in 1975. Herb had a world-class wine cellar and, on occasion, his wife Helen would be the one to have to open a bottle of wine for her guests. She had a hard time with the task at hand and finally presented her husband with the engineering challenge of a lifetime. "Produce a corkscrew that would extract the cork effortlessly and perfectly every time."

The Screwpull Table Model evolved from this challenge, it was able to remove a cork with very little force, inflict no injury, and perform without embarrassment. The Screwpull was an instant success—it was a beautifully designed corkscrew that ensured removal of the cork without pulling.

The Screwpull maintains a place in the New York Museum of Modern Art for its excellence in design.

Herb Allen later developed the Screwpull Pocket Model and then the Screwpull Lever Model. The Screwpull family is truly the only major innovation or invention in corkscrews in the last 100 years. October 10th, 1981, in the Council Chambers of Horsham Town Hall, Horsham, West Sussex, England, Herb Allen was officially welcomed into membership of the International Correspondence of Corkscrew Addicts as its first honorary member for his contribution to corkscrews, by invention.

Acknowledgments

Much of the information you find in this book comes from my one-on-one, ten-year-plus association with the late Herb Allen himself as he developed, perfected, and then had to protect his corkscrews and patents. As a corkscrew collector, working with Herb was a priceless experience. His engineering mind looked at corkscrews with a totally different perspective than through the eyes of a collector.

The late Virginia Bedford was Herb's Executive Secretary. If you knew Herb you knew Virginia. Not long after Herb died, Virginia called and asked me to come by her office. She had organized a large folder of information about Herb from his early days to 1990. She said, "Don, someday you are the one that will have to write the book on Mr. Allen and the Screwpull. You know everyone here, your connection to the ICCA, your knowledge about corkscrews and your collection of the prototypes are all important in writing the book." At the time I didn't realize that anything would ever come of her statements. Time has passed and here we are. The book is a reality.

There are many others who provided items and information to make this book possible. Listed below, in no particular order, are some of those gracious people:

Herb Allen's family: Anne Symonds and Michael
Loretta Haren, VP Hallen Company US
Vickey Kaufman, Director Hallen Company, Ltd. UK
The late Roland Pair. Herb made many of the early prototypes himself, but Roland was a machinist who made most of his more complex devices.
Margaret Anderson, patent attorney for Mr. Allen
The late Bob Roy, Richard's Liquor Store
J.J. Krchnak, photographer responsible for some of the early photos
Karel Janousek, LeCreuset Canada
Monica Pinto, LeCreuset France
Don Bull, ICCA, Mirth Right
Rice University Fondren Library, Woodson Research Center, Special Collections Archives

I would like to thank you all, including some who are not listed, for your contributions.

Introduction to Screwpull, Houston, Texas, and Rice University

The incubation period for the development of Screwpull and Leverpull was the years 1974 to 1978. Somewhere around the first part of January, 1979, Screwpull was born. This took place in the home basement workshop of respected and talented engineer, Herbert A. Allen. The carpeted workshop was fitted out with the latest milling machines, lathes, drill presses, grinders, and bin after bin stocked full of nuts, bolts, rivets, and many other parts and pieces—a machinists dream. The workshop adjoined his world-class wine cellar, separated only by a wall and a few steps. Herb's already brilliant mind was enhanced and molded by his studies at Rice Institute in Houston, Texas, Class of 1929. Herb's connection with Rice lasted for the rest of his life, as well as thereafter.

William Marsh Rice's murder was one of the most sensational cases of the 20th century. It certainly created a stir in 1900, when the incident and the subsequent trial made international headlines. Rice's biographer, Andrew Forest Muir, called the sordid New York City incident "Murder on Madison Avenue."

If not for Rice's attorney, Capt. James A. Baker (grandfather of James A. Baker, Secretary of State under President George H. W. Bush), who was suspicious of the details surrounding Rice's death, the five million dollar charter for establishing William M. Rice Institute (name changed to Rice University in 1960) would have been swindled away by chloroform-wielding cohorts. Rice built the Bayou City (Houston, Texas) with everything from plank roads, railroads and steamships to cotton industry investments and real estate developments.

Rice came to Houston in 1839, eyeing the town as a land of opportunity despite jokes in his Massachusetts hometown newspaper about how Houston's currency consisted of cows for large sums and calves for change.

Much of Rice's fortune was accumulated through the Rice and Nichols importing company. The firm offered a novel service, banking for its customers, because the practice was illegal in Texas until 1870.

Over the years Rice transferred money from his estate into a fund for the Institute. Several associates, including Baker, knew of Rice's wishes to build a university after his death. Unfortunately, Rice's $55-a-week secretary and a lawyer who had been working against Rice in the contestment of his second wife's will, had something else in mind for that money.

Charles Jones and Albert T. Patrick concocted what might have been the perfect murder plot had they not botched events following the actual killing. Baker recognized foul play when he spotted a check Jones forged incorrectly. Jones accidentally omitted the letter "L".

When the murder case finally ended, the already appointed trustees of Rice Institute set plans in motion to actually build the university. The group recruited Edgar Odell Lovett, an astronomy professor at Princeton University, as president. He established a curriculum, oversaw the acquisition of 277 acres on Main Street and hired a faculty.

Rice University and Lovett Hall, with its giant Sally-Port arch, still stand as monuments to the man who had an educational vision and enough sense to associate with at least a few honest business leaders who would follow through with his wishes ("Lovett Hall, Rice University – 6100 Main Street," by Danni Sabota, *The Houston Business Journal*, September 16, 1991).

Lovett Hall, Rice University, Houston, Texas, completed in 1911

Herb Allen and friends in his 1920s Mercer Speedster, c.1929. *Courtesy of 7*

Chapter 1: Herbert A. Allen – The Inventor

May 2, 1907 – June 12, 1990

Herbert A. Allen, c.1970.
Courtesy of Anne Symonds, Bert Brandt, photographer

Herbert A. Allen was born in Ratcliff, Texas. He received a B.S. degree in Mechanical Engineering from Rice University in 1929. He spent the next two years in miscellaneous research. From 1931 to 1935 he served as Chief Engineer for the Abercrombie Pump Company in Houston, Texas. From 1935 to 1977 he served as Chief Engineer, Vice President, President, and Chairman of the Board of Cameron Iron Works, Inc. Houston and from 1977 to 1989 he was Director of Cameron Iron Works. From 1977 to 1990 he also was Chairman of Hallen International.

He held many outside directorships as well, some of which were Big Three Industries, Texas Commerce Bank, Tenneco, Inc., Houston Symphony, Houston Chamber of Commerce, William Marsh Rice University, Museum of Fine Arts (Houston), and Texas Tech University. Some of his professional memberships included the American Society of Mechanical Engineers (ASME), American Institute of Mining, American Petroleum Institute (API), Texas Society of Professional Engineers, and the Houston Engineering and Scientific Society. Herb was a registered Professional Engineer in the state of Texas.

Herb Allen met Helen Daniels in 1936, on a date arranged by friends, and it was love at first sight. They were married in 1937 and they lived in Houston where they raised three children: David, Michael, and Anne (Symonds).

Herb Allen operating an engine lathe in the Mechanical Engineering Lab, Rice University, c. 1926. *Courtesy of 7*

Herb Allen family photo, c. 1990 (L to R): Herb's wife Helen, Herb, son Michael, and daughter Anne (Symonds). Herb's name tag reads: "OTC '90, Ocean Technology Conference 1990, Houston, Herbert Allen, Exhibitor." Herb never stopped, still working on his last days. *Courtesy of 7*

Herb held more than 380 Oilfield Related Patents, U.S. and foreign, as well as the Screwpull® family of patents. At the time of his death, he had 56 patent applications pending. Herb Allen's record of accomplishments speaks for itself. The photo above was one of the last photos taken of Herb Allen.

Herbert Allen, of Cameron Iron Works, was presented the Distinguished Achievement Award at the 1990 Ocean Technology Conference (OTC) in Houston, Texas, "For pioneering achievements as inventor and engineer in the development of well control equipment for offshore drilling and production operations world wide; in recognition of more than 300 patents covering assemblies for blowout preventers, pressure relief valves, and chokes; and for his inspirational corporate leadership."

There have been many stories told and published about his climb to success. These included his struggle through the Great Depression of the 1930s and his ingenuity during the war years of the 1940s in converting Cameron Iron Works into one of the major suppliers of wartime materials. He received the Navy's "E" award of excellence and contribution of effort. He successfully modified the construction of depth charge projectors, eliminating firing malfunctions; designed and set up an automated plant to produce the parts used in firing depth charges; developed ways of manufacturing 3-inch and 50-caliber naval gun barrels; designed hydraulically operated mounts for 40-mm Olerikon rapid-fire guns; and developed the equipment, alloys, and procedures for making superior breech blocks. His inquiring mind, always thinking about how "there is a simpler better way of doing it," successfully graced him with the many patents, honors, and awards he received for his genius in engineering oilfield equipment and corkscrews.

Herb Allen, lower far right with folded arms, in his early years of solving problems in the oilfields. A crew of roughnecks on an oil drilling platform, c.1930s. *Courtesy of 7*

An oil field pump and its creator, Herbert Allen, c.1940s. *Courtesy of 7*

Many of us would not understand what most of his oilfield patents entail. Pumps, valves, down-hole devices, and hundreds more have all contributed to the high level of technology that the oil and gas industry enjoys today. Several of his patents that most of us do understand are the greatest inventions ever devised for the safety of the oilfield roughneck. The oil and gas well blowout preventers he designed and perfected have contributed significantly to the welfare of the public. These devices have been credited with saving hundreds of lives on the oil and gas field drilling rigs of the world and conserving many man hours of time and countless quantities of oil and gas. Save for Herb Allen's efforts, many people would have been lost.

Cameron Iron Works had its start in 1920 as an oil tool repair shop in Houston, Texas. Herb Allen came on board as a young engineer in 1931. The three partners of Cameron were Harry S. Cameron, Edmond L. Lorehn, and James S. Abercrombie. The last surviving partner was Abercrombie, but in 1950 he could no longer stay active in the day-to-day management of the company. From 1950 until his retirement, Herb Allen became the guiding hand of Cameron Iron Works. They maintained their oilfield products but also became a world leader in specialty metal forgings. In 1967, the first supersonic Concorde passenger plane lifted off with its powerful jet turbine engines turning on special metal shafts and other moving engine parts that were put to extreme loads and pressures. The shafts were developed and manufactured by Herb and his Cameron team. They developed nozzles for the Titan space booster and missile cases for the United States Government military. Herb Allen was truly an outstanding pillar of his community and his country.

Herbert Allen in the oil field among wooden derricks, c. late 1940s. *Courtesy of 7*

More oil field devices and their creator, Herb Allen, c.1940s and 1950s. *Courtesy of 7* (above and right)

Herbert Allen approached his Rice University activities with the same energy and enthusiasm characteristic of his role at Cameron Iron Works. He was named to the Rice Board of Governors in 1949 and became Chairman of the Board of Trustees in 1972. To Mr. Allen, one of the most interesting of his Rice projects involved planning for the Rice Stadium in 1950. "When the decision was made to build the stadium I took a few months off and worked with Gus Wortham, the trustee in charge of the project. I wanted to build a stadium where you could get in and instantly see the game. I still get a thrill out of that stadium every time I drive by it. It's the finest college football stadium in the world." As a trustee Allen was also involved in the establishment of the college system at Rice in the mid-1950s ("Rice Recognizes Herbert Allen High Emprise," Nov–Dec 1974).

Rice Stadium, indeed, is a fine stadium and was the host of Super Bowl VIII. There are very few colleges that can boast of having their stadium facility used as the site of a Super Bowl. Rice University and Houston, Texas, are lucky to have had trustees with the enthusiasm and love for football that resulted in this stadium. In 1988, the *Houston Post* ran a story on the late Bill Kirkland, another prominent Houston figure, that included the following paragraph:

A varsity baseball pitcher at Princeton, Kirkland was scouted by the New York Giants before he opted for banking as a career. But the love of baseball never left him. He was chairman of the Harris County Board of Park Commissioners, the quasi-political force behind the birth of the Astrodome in Houston. Kirkland used to amuse his friends by telling of the first sketch of the domed stadium being drawn on a paper napkin by the intrepid and imaginative Herb Allen. Kirkland and Allen, along with R. E. Smith, George Kirksey, Reuben Askanase, Craig Cullinan, George Bruce and Aaron Farfel were the movers and shakers at the vanguard of the movement to bring big-league baseball to Houston in the '50s ("Bill Kirkland, Last of the Country Gentlemen," by Charles P. Thobae, *The Houston Post*, July 9, 1988).

Herb Allen, left, not wasting time on a coffee break. His slide rule was always at hand, calculating a better way of doing things, c.1950s. *Courtesy of 7*

Herb often said: "I've always counted myself one of the luckiest people who ever lived. Finding a job with a growth company during the depression and being in the right place at the right time with an 'idea' to fix something or invent a better way of doing it. The thing that really counts in life, I think, is a sense of some internal satisfaction—a feeling that you've done something that you yourself are pleased with. It doesn't make much difference what anyone else thinks."

Herb often used the following quote in varying ways in his acceptance speeches for his many awards:

> *For me, invention—if you want to call it that—has always been a straightforward process. A very important part of the process is to clearly set out a need. I'll think about the problem until I get tired of it, and then I'll forget it. After a while—days or months—something triggers a new idea or approach. I'll say to myself, "why in the devil didn't I think of that before?" You don't always solve the problems you are trying to solve, of course, but if you are fortunate, you get some answers that may work somewhere else.*

Early in 1979 I studied an advertisement in the local newspaper, run by a local liquor store chain, depicting a new type "Teflon" coated corkscrew. Since the owner of that chain was a client of mine, I decided to go over and find out what I could from him about this new corkscrew. ("I" being a newly inducted member into the International Correspondence of Corkscrew Addicts). It was exciting, to a corkscrew collector, to watch Richard Trabulsi and his assistant store manager, Bob Roy, demonstrate this new device, the ease of the operation, their excitement with the product, the acceptance by the public as evidenced by their sales and I bought the last one on the shelf.

As we were talking, Michael Allen, Herb Allen's son, drove up in a Mercedes-Benz convertible and brought in a new "shipment" of Screwpulls. (Richard's Liquor Store, with a great wine selection, was the first and trial market for the Screwpull).

We were introduced and after some conversation it was obvious that Michael had a keen interest in the mere existence of an organization called the International Correspondence of Corkscrew Addicts.

Two hours later, in my office, I received a call from Herb Allen inviting my wife, Betty, and me over for a glass of wine in his wine cellar. We accepted and it turned out to be a wonderful afternoon and the beginning of a new relationship we will never forget. Herb, along with his wife Helen, son Michael, and daughter Anne Symonds were present. We tasted some exquisite wines as the result of several Screwpull demonstrations. Herb took me on a tour of his basement workshop where he built his prototypes. The workshop contained tools and machinery of the latest technology. It was a very impressive workshop with the capability of producing most anything and an inspirational environment in which to do it. This is where Herb exposed his first prototype of the Leverpull to me and explained some of the problems he was having with it.

Later that year I took several new Screwpulls to the Annual General Meeting of the International Correspondence of Corkscrew Addicts. On September 24, 1979, the Right's Reception (The ICCA has a Right instead of a President) was held at the Garrick Club in London. Just Right Brother Timothy Diener (Cellar-master of Christian Brothers Winery) demonstrated one of the Screwpulls for us on a select group of Christian Brothers wines brought from Napa, California. After the excitement of the demonstration settled down, and many technical questions were being asked, I reported to the correspondence as much as I knew about it as follows:

> **Body and Handle**: D.E. Polycarbonate Plastic (LEXAN) with the strength of mild steel but with a modulus of elasticity 1/8th that of steel (will flex 8 times as much as steel)
>
> **Screw**: The wire screw is .105 inches in diameter, high strength steel, changing soon to .100 inches in diameter using #302 stainless steel with a tensile strength of 200,000 pounds per square inch, and is Teflon coated. The point of the screw is ground on the topside of the wire so that the lead of the screw pulls up, rather than down or out.
>
> **Stand**: A solid color plastic that will be changing soon to clear plastic so that the full length of the screw is visible.
>
> **Body and Handle Colors:** The colors presently are burgundy, amber, white, and black. Other colors are available and will be brought into the market a few at a time.

As one of the newest members, I was extremely proud to be able to bring this information straight from the inventor to the most learned assemblage of corkscrew collectors in the world. During one of our conversations, Herb said, "I have never really had trouble opening a bottle of wine with the available corkscrews, but after watching other people and the troubles they encountered with the commitment, I decided I had to do something about it. Especially for the ladies—most of them look at a corkscrew as if it was a two-headed snake and really do not want to tackle the task at hand, (obviously recalling memories of past experiences), plus I had the challenge presented to me by my wife Helen. I feel like I have taken away the fear of pulling a cork with the Screwpull."

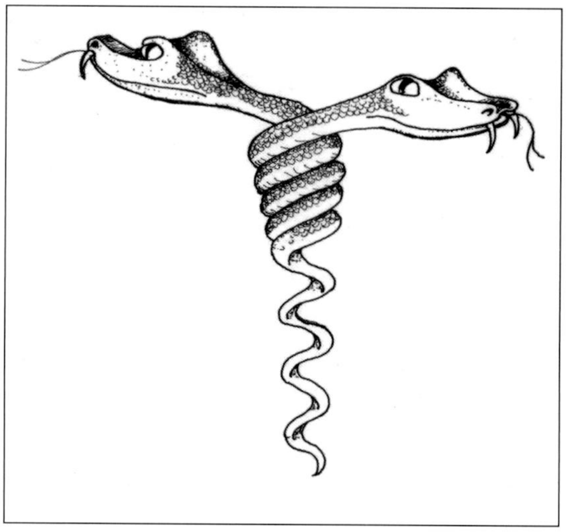

Two-headed snake. *Courtesy of DM.*

Herb had an idea of how to make a better corkscrew. He had all of the technical skills, know-how, and means to produce the idea. To implement this idea, in the best way possible, he spent many hours with his friend, Bob Roy, wine expert and Vice President of Richards liquor store. They discussed "what a corkscrew should be." How long should the screw be? Should it fully penetrate the cork? What should we call it? Bob said that these meetings rarely drifted into technical discussions because Herb already knew what he wanted to do and how to do it. "Our discussions were about design, consumer appeal, and how the product would be received by the wine enthusiasts of the world. After prototypes of the "new idea" started evolving, Herb spent many more hours with Bob testing them on all types and sizes of bottles and corks. "The engineering was already done. We were testing them for flaws or if we could see ways to improve them."

Herb Allen's idea was to use a friction-reducing coating or material embedment on a self pulling wire screw. The screw would be made of high tensile stainless steel, allowing the use of a smaller diameter (gauge) wire than had been used in the past, reducing the possibility of cork breakage. POLYTETRAFLUOROETHYLENE (PTFE) is the plastic resin, manufactured by DuPont, that was used in industry for many friction-reducing applications. In 1959, pure PTFE coatings were sold as FLUON. In the 60s and 70s it was commonly known as TEFLON-"S", providing the term TEFLON as it has come to be referred to in everyday language.

In time, the PTFE outer coating would wear off of the first several coils of the helix nearest the point. A heavily used screw would eventually glow shiny silver at the worn areas of the stainless steel screw. Some consumers thought that their corkscrew was worn out when this occurred, but in fact was not. PTFE particles are imbedded into the steel during the coating process as well as on the surface of the steel so that the wearing off of some of the coating does not significantly affect the friction reducing capabilities of the PTFE.

Very early on in production, Herb switched from PTFE to a newer type TEFLON coating, also produced by DuPont, called Silverstone SUPRA. He considered Silverstone SUPRA to be harder and more durable. Early production screws were coated in California. After some time, the coating was applied in London and later in Houston, Texas.

Driving a coated wire helix through hundreds of hard corks and extracting it is a very severe test for any coating such as this, so much so that DuPont started using the Screwpull screws for testing their newest friction reducing coatings.

Herb ran many engineering calculations and made many prototype screws in his effort to come up with the "perfect" screw. The diameter of the wire, the diameter of the helix, the pitch of the screw, the friction reducing contribution, and forces the screw would have to withstand, all became part of his formulations and his trial and error methods. His tireless efforts indeed have created a near "perfect" screw.

Herb made many speeches in his career as CEO of a major energy company, Chairman of the Board of Rice University, engineering societies, etc. Many times, during these speeches, he would drift away from the oilfield and start in on corkscrews and get involved in telling the public about the different aspects of how a cork should be pulled from a bottle of wine and why it should be done in a certain way.

Herb Allen in his wine cellar opening a bottle of Montrachet with a Table Model Screwpull, c. 1988. *Courtesy of J.J. Krchnak.*

The following excerpts are from a paper Herb Allen wrote in 1979, called: *The "Screwpull" - EVERYBODY'S CORK PULLER.*

> *Nearly all wine lovers have a collection of cork pullers that represent both their search for a satisfactory design and gifts from friends who think they have found the answer. The very fact that such a variety exists is reasonable proof that no one design has proven greatly superior in the only thing that it is intended to do—infallibly pull the cork. When serious thought was directed to the reason for frequent failure, it was determined that the degree of skill is very important—almost any device offered will achieve satisfactory results in the hands of an expert. The necessity for correct insertion of the screw has, up until now, called for expertise well beyond the capability of most.*
>
> *We watch with fascination as the sommelier takes the folded waiter's screw from his pocket unfolds it and expertly applies it to remove the aged cork, completely whole and undamaged, from the fine old bottle and offers it for our inspection. There is no question but that it is the sommelier's skill, gleaned from instruction and from thousands of such acts that make for such success. You have also seen the same gadget destroy the cork when experience and skill are not present.*

This brings to mind an event I will never forget. While enjoying a great string quartet from Romania, I could not help but notice the beautiful sweet tone of the first violin. During a break, I mentioned to the second violin player, who was his daughter, that he has such a beautiful sounding violin. Her reply was: "Yes, it is a nice old violin, but the 'hand' is what really makes the difference."

> *Most of the innovation of some two hundred years of cork pulling devices has revolved around reducing the force necessary to pull the cork. In all too many designs, this very reduction in the force required has resulted in faster cork movement with disastrous results, and that is why the various designs that employ a screw rather than a linkage are, on the whole, more often successful.*
>
> *Some corks may require a pulling force of 100 pounds or more, although a more normal figure is 50 pounds or less to start movement. Normally, the moving force remains relatively constant for much of the cork length since most of the drag comes from the lower end of the cork that is moist and tends to retain a tight fit in the bottle. The high variation in pulling force is a reflection of the cork's inherent physical properties, length and condition.*
>
> *Of all the screw configurations that have been tried in the past, there is almost complete unanimity in the view that the helix is preferred. Wire diameter, pitch diameter, and lead have varied, but again, it is almost self-evident that minimum rupture stress will result in the cork if the point of the screw is in line with the helix; the wire is of minimum diameter and the lead is such that when fully penetrating the cork, enough turns will be engaged to always pull the cork and not tear out of soft, weak corks.*
>
> *The use of a guide frame and a single long screw fully complied with the requirement that the pull be in line; a single element provided both. The reduction in force to penetrate the cork also answered the requirement that minimum force be required in operation. Only 3 to 5 inch pounds of torque are required to penetrate the cork, and twice that to effect removal.*
>
> *The "Screwpull" answers the need for a completely reliable cork puller that can be easily operated by someone pulling a cork for the first time.*

With the Screwpull, the "hand" is not all that important. The Screwpull and Leverpull brought great rewards to the Hallen Company from sales from all over the world. Herb Allen continued with improvements, developing related items such as the champagne star cork remover, the champagne re-sealer, the foil cutter, etc. until he was over eighty years old. Herb died on June 12, 1990, at the age of 83.

The Hallen Company was in operation from 1978 until 1991. Herbert Allen was president and chairman of the board over all operations. The Board of Directors also included Anne (Allen) Symonds and Michael Allen, and Virginia Bedford was treasurer. Loretta Haren was vice president of Hallen Company HOUSTON and Vicky Kaufman was director of Hallen Company, Ltd. (UK branch of Hallen Company US).

Left to Right: Herbert Allen (ICCA Honorary Member), Hous-RIGHT Donald Minzenmayer (ICCA Member - Author), and How-RIGHT Perry Howland (ICCA Member) discussing the Screwpull and the Leverpull at the Right's Reception of the International Correspondence of Corkscrew Addicts' Annual General Meeting in London, 1984. *Courtesy of Herb Danziger*

Herb's close connection to Rice University, as a graduate and chairman of the board, prompted him to bequeath the Hallen International Company all of the Screwpull patents, logos, and name recognition to Rice University in his will. Rice considered this a valuable donation to the University and treated it with respect. Rice University, being an institution of higher learning, was not in the manufacturing business, so it put Hallen International up for auction. LeCreuset, a well-known kitchenware company headquartered in France, presented the high bid. LeCreuset has continued making improvements and additions to the Screwpull family of cork removers and Screwpull products. In my opinion, Herb would be proud of the improvements and new products that LeCreuset has added to his beloved Screwpull line of products.

Screwpull has become the generic name used for a resin-coated self-pulling wire corkscrew no matter where it was made or who or what company made it, much like the term "XEROX machine" is used to describe copy machines no matter what brand they are.

Sometimes it is the little things that brighten our days. A corkscrew that never fails is one of those little things that gives us a lift when it seems like nothing else will work.

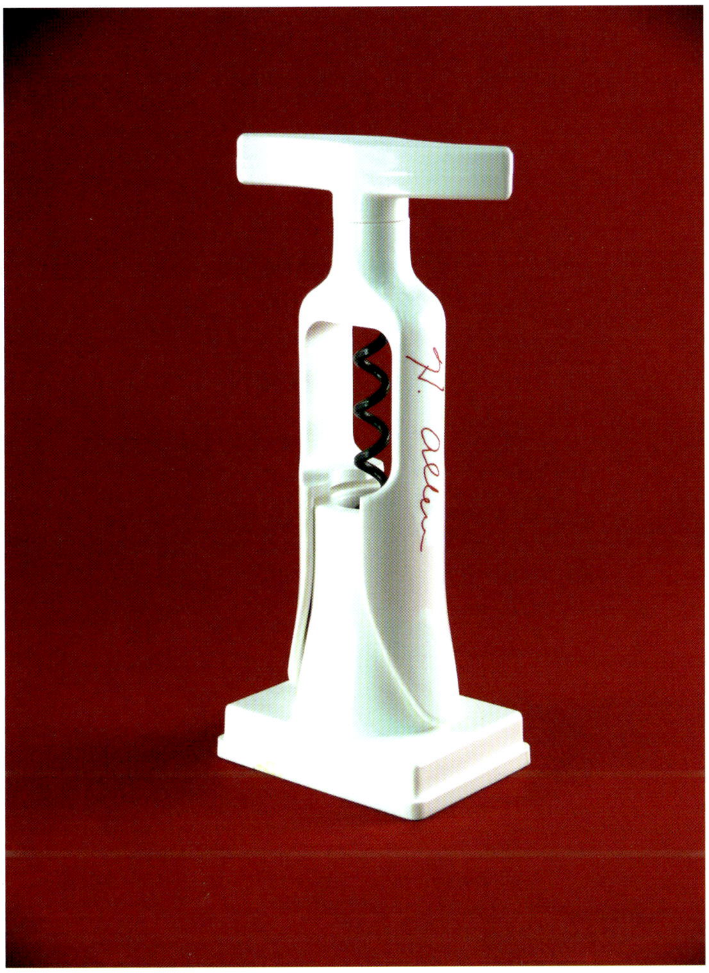

A very special Version 1 Table Model Screwpull. *Courtesy of Loretta Haren.*

The ICCA logo. *Courtesy of DM.*

The ICCA, The International Correspondence of Corkscrew Addicts, has been mentioned previously and will be mentioned again, so at this point I will briefly explain what that is. The ICCA is the most prestigious corkscrew collector organization in the world. The members are a very diverse group, including doctors, architects, attorneys, engineers, investment bankers, farmers, and many other occupations. They come from three continents, all with the same addiction, collecting corkscrews. Their homes are filled with world-class corkscrew collections and they continue to find and research corkscrews of all types.

The first meeting of the ICCA was held October 1, 1974, organized and hosted by Dr. Bernard Watney and held at the Guinness Brewery in London, England. It was decreed at this meeting that the leader of this group, rather than being known as the president would have the designation *Right*. Dr. Homer Babbidge, president of the Hartford Graduate Center in Hartford, Connecticut, and a founding member, suggested at the meeting that in the words of American Congressman and Statesman Henry Clay, that it was better to be Right than president. On that theory, the ICCA leader has hence been known as the Right. Brother Timothy Diener was elected the first Right and was also designated as the ICCA's Chaplain. Brother Tim was a priest, the cellar master of Christian Brothers Winery, and also a founding member of the ICCA. Dr. Watney, chief medical officer of the Guinness brewery, was elected to the Chief Correspondent's post, the only other elected office.

At the end of the Right's term of office, he is designated an honorary title. From 1974 thru 2011 the Past-Rights and their titles are as follows: Brother Timothy Just-RIGHT, Dr. Bernard Watney Start-RIGHT, Dr. Homer Babbidge All-RIGHT, Don Morway More-RIGHT, Bob Nugent Nu-RIGHT, Perry Howland How-RIGHT, Richard Dennis Spend-RIGHT, Dr. David Bradshaw Try-RIGHT, Don Minzenmayer Hous-RIGHT, Don Bull Mirth-RIGHT, Ron MacLean Can(e)-RIGHT, Joseph Paradi Pro-RIGHT, Helgir Solheim Sol-RIGHT, Wolfgang Handel Hand-RIGHT, Dr. Bert Giulian Read-RIGHT, and Fred Kinkaid is Left-RIGHT.

The ICCA membership is fixed at fifty members worldwide. The membership includes addicts from Australia, Austria, Canada, Denmark, England, France, Germany, Greece, Holland, Italy, Netherlands, Norway, Portugal, Spain, Sweden, Switzerland, the United States, and Texas.

The word "Correspondence" in the name of the society means that the members correspond. Prior to e-mail and cell phones, we used the mail system and expensive telephone calls. We, members and their Go-withs, meet somewhere in the world for our Annual General Meeting for a week of corkscrews (trades, dinners, auctions, talks, tours, etc.), hosted by one of the members. Yes, it has been a wonderful excuse to see the world with a fantastic array of friends.

The ICCA badge

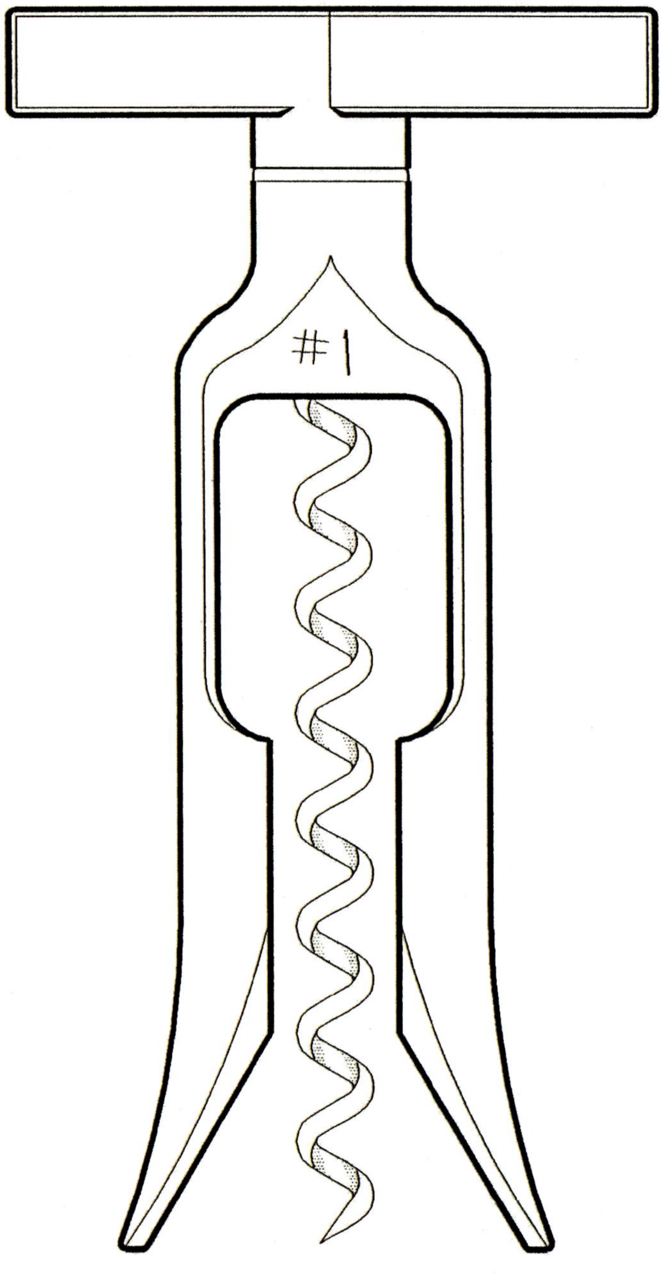

Drawing of Pre Version 1, #1, prototype. *Courtesy of DM.*

Chapter 2: Screwpull Prototypes

Screwpull is the registered trademark of a type of corkscrew that was designed so any man or woman, weak or strong, could remove the cork from a wine bottle with very little effort and not injure themselves in the process. There have been thousands of designs in the past 250 years that have tried to accomplish a product with these criteria, most of which were just improvements or variations of what was already there. The types of screws have changed very little. The bladed center worm, the pointed center worm, the Archimedean worm (sometimes known as the auger worm), and the wire helix have all been in use since about 1750. The mechanisms are what varies greatly, from direct-pull T-handle corkscrews (probably wining the prize for the most manufactured) to complex mechanical devises. Through the years, the wire helix has been proven superior to the others, particularly in the area of preventing cork damage.

The Screwpull falls into the category of "a self-puller." The self-puller is not new as evidenced by the following photographs. A self-puller has a clockwise rotation as the screw penetrates the cork and the downward movement of the screw is arrested by a frame or collar. Continued rotation causes the cork to ride up the screw, out of the bottle, and into the frame. So, what is new? The major breakthrough was Herb Allen's idea to reduce the friction of the screw and using a smaller diameter wire to help prevent cork breakage. This was all made possible by using a long, high-strength stainless steel wire helix with an anti-friction coating. The coating reduced the turning forces required by a factor of more than six times.

Allen experimented with many types of frames to incorporate his idea and came up with Lexan plastic material as his choice. Lexan is extremely strong, with the strength of mild steel, yet flexible. His high tech idea of removing a cork from a wine bottle is what the world had been waiting for as evidenced by their eventual sales. This product solved many of the problems encountered with cork removal and eliminated many embarrassing situations.

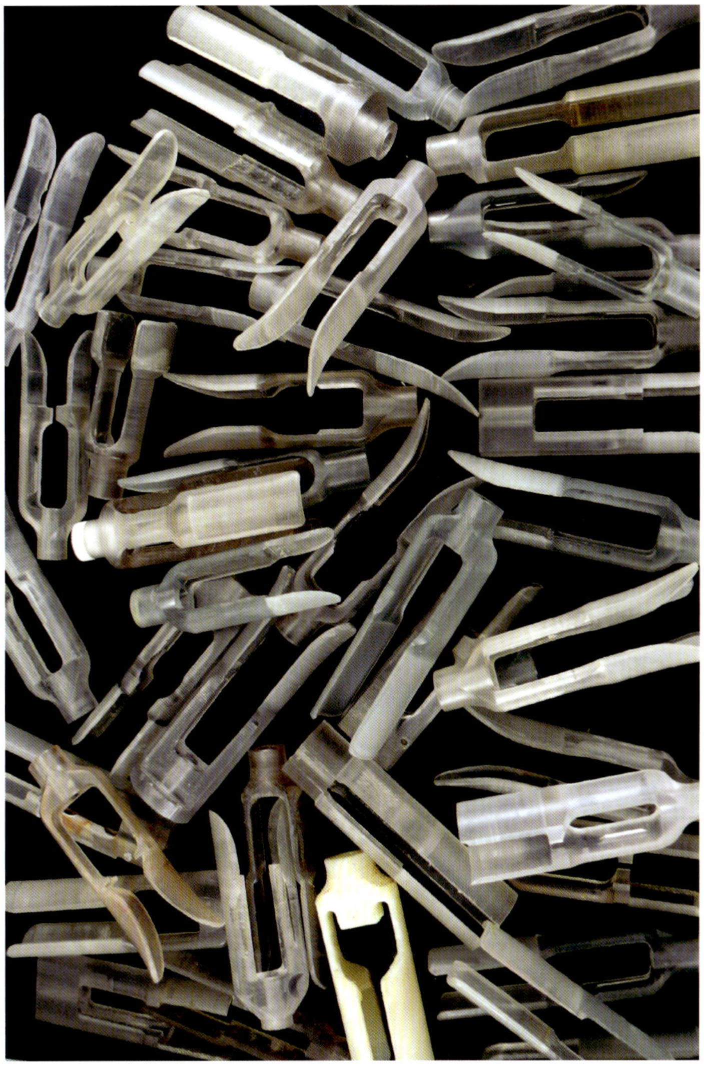

Screwpull prototype rejects.

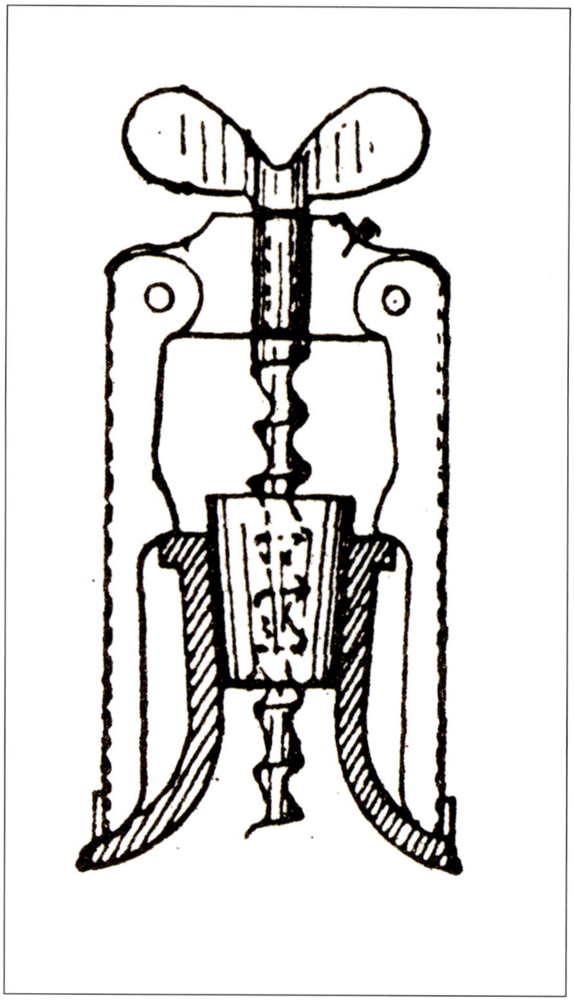

A self-puller with a bladed center worm. There are several of these known to exist. It was patented in 1897 in England and America by C. Blombach and manufactured by Hollweg.

An early- to mid-eighteenth-century self-puller with an Archimedean screw.

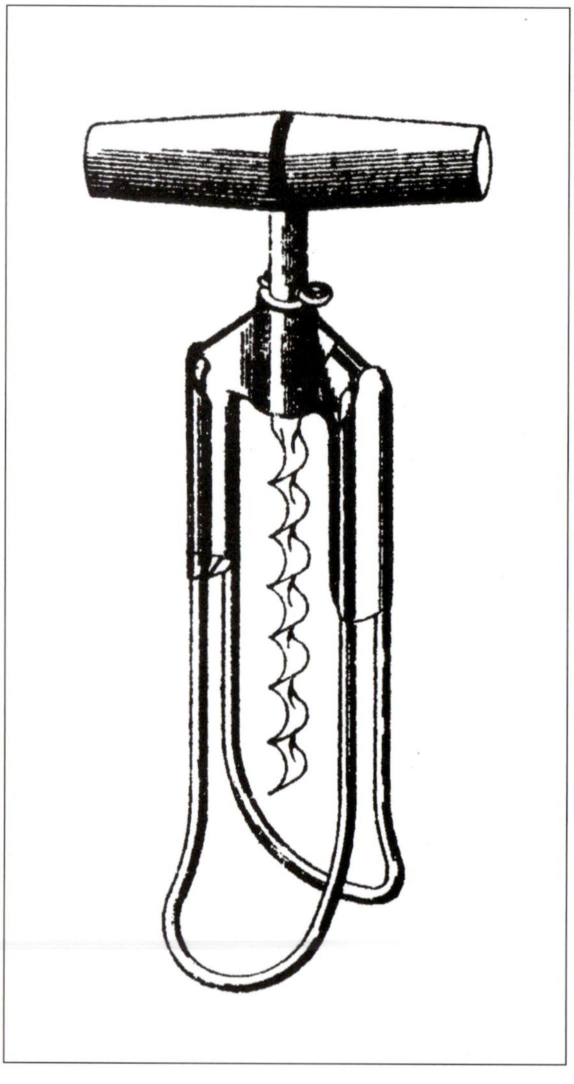

A self-puller with a bladed center worm and a frame with wire legs. It was patented by A. Baumgarten in England and America in 1901 and may have never been manufactured.

Prototypes

Herb Allen's idea, and how he wanted to do it, came together fairly quickly, but working the bugs out consumed many hours of his time. He made many prototype frames to solve design problems and many screws were made to solve engineering problems.

It is interesting to note the differences in the screws. They are all 5 inches long and the same 3/8" (9.5 mm) outside diameter but they vary in pitch and coating. Pre-version #1 and Prototypes #1, #2, and #4 have a pitch of 15 mm (distance to complete one full spiral of the screw) with a white friction reducing coating (plastic resin commonly called TEFLON) manufactured by DuPont. They all have a high tensile stainless steel wire screw, with a wire thickness of approximately 2.6 mm, except Version #1 Prototype #3, which is approximately 2.35 mm in diameter. Version #1 Prototype #3 has a pitch of only 12 mm and a black coating. It may seem that #3 had the wrong screw in it because the body is almost identical to #1, #2, and #4, except the white screws from the other prototypes will not go into the bushing of #3, and for some reason the handle is longer (possibly just an experiment to see if more leverage was necessary). After measurements with a digital caliper, it is hard to see a measurable difference in the outside diameter of the black screw of #3 in relation to the white screws of the other early prototypes, but they just will not go into the bushing of #3. However, the gauge (or thickness) of the white screws was 2.6 mm, where the #3 black screw had a gauge of 2.35 mm. If the screws were made in the same way, it would mean that the outside diameter of the black screw would be approximately .2 to .25 mm smaller than the white ones. That fraction of a millimeter made a difference. Herb Allen, being the engineer that he was, would know this difference, and, not wanting a sloppy fit for this experimental screw, apparently made a special bushing the right size for Prototype #3. Prototype #5 has a pitch of 15 mm with a tan-colored coating.

Screwpull Prototype #1, (pre-Version 1 Screwpull), marked "#1" both sides and on handle, signed, "H. Allen May 23, 1978" on one leg and "H.A." on the handle. The markings are hand engraved with a vibrating marking tool, except one of the #1's, which is stamped. Note the white bushing in the top of the frame.

Screwpull Prototype #2 of the pre-Version 1 Screwpull, marked "2," hand engraved on one side. Note the white bushing in the top, c. 1978.

Screwpull Prototype #4 of the pre-Version 1 Screwpull, unmarked, c. 1978. *Courtesy of Try-RIGHT Dr. David Bradshaw.*

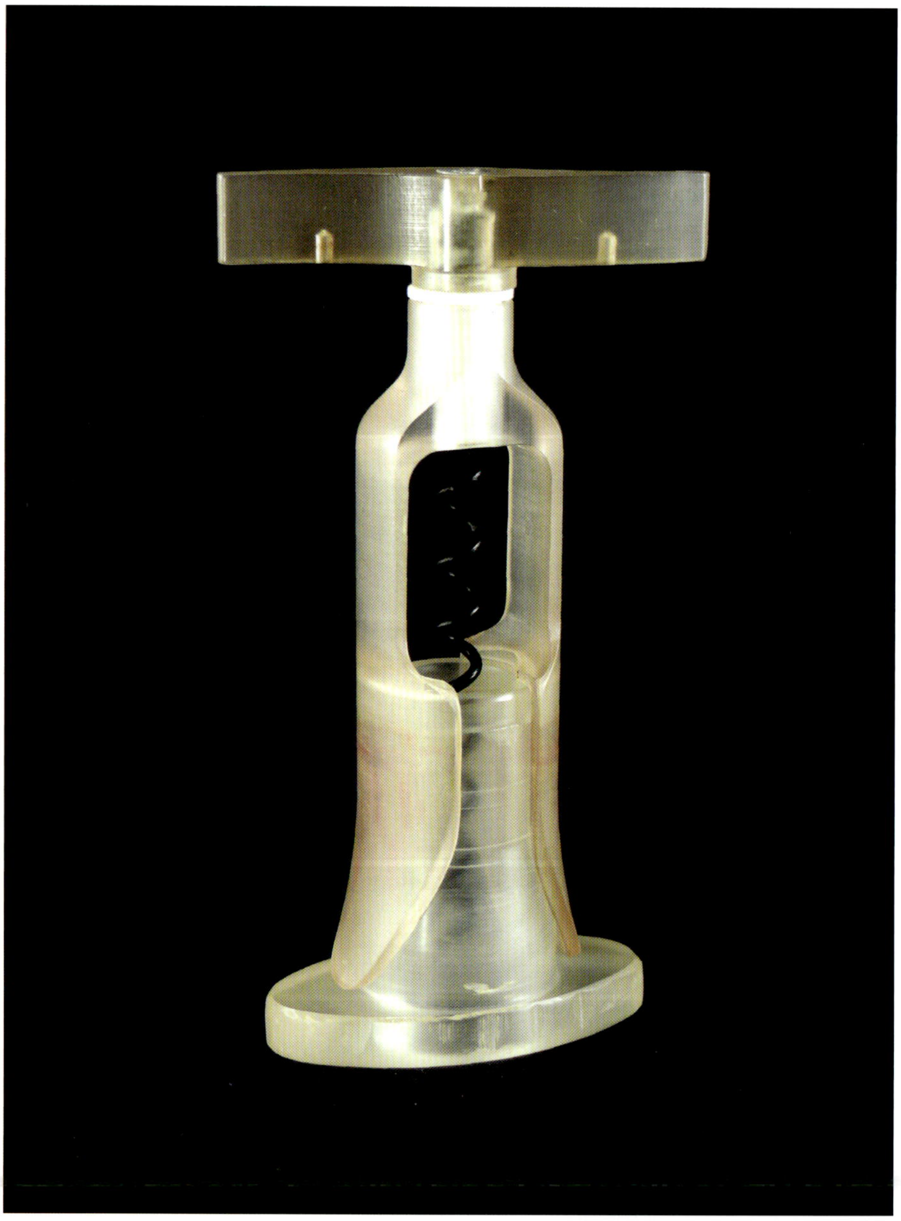

Screwpull Prototype #3 of Version 1 Screwpull, marked "3," hand engraved on one side. Note the black screw and the white bushing in the top, c. 1978. This prototype is believed to be #3 of the Version 1 Screwpull prototypes, not the Pre-Version 1 prototypes.

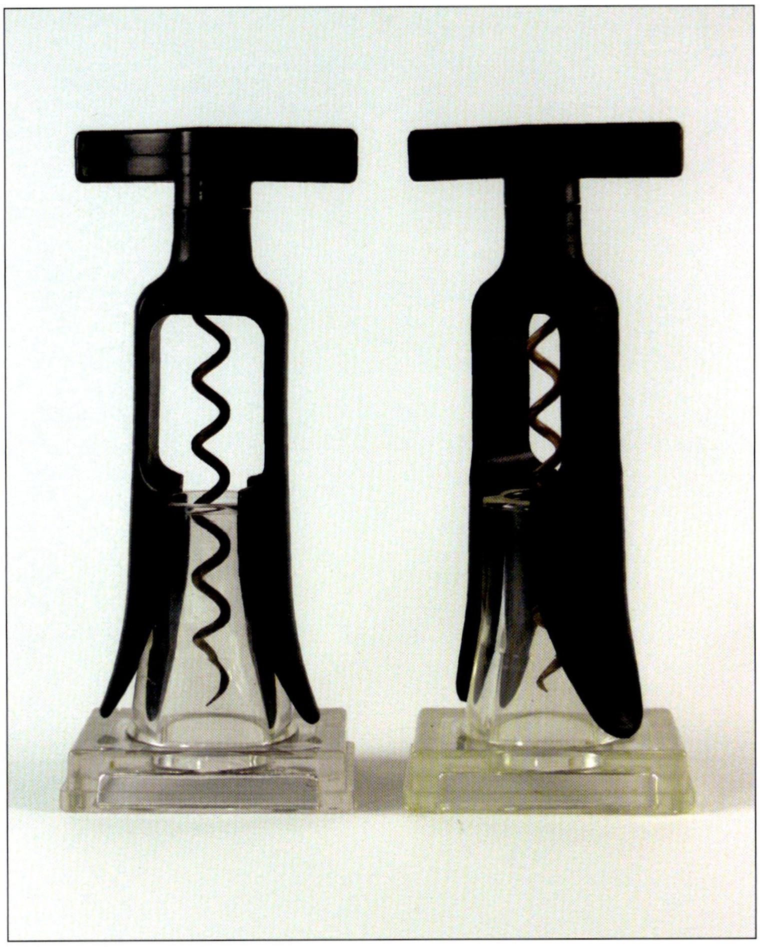

Version 1 Screwpull Prototype black composite material, unmarked without Screwpull logo, c. 1978, the last Version 1 Screwpull prototypes prior to production.

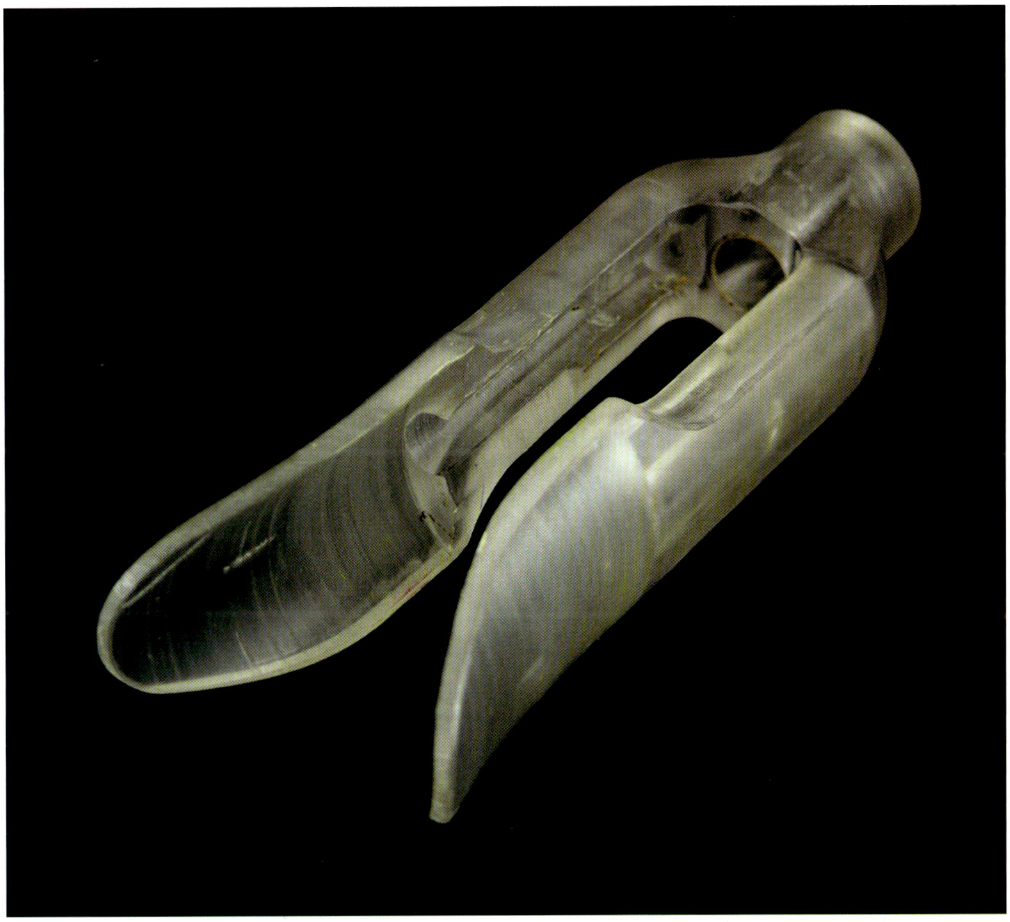

A Version 1 Prototype where Allen was experimenting with a slot up the inside of the upper part of each leg of the frame to prevent the cork from turning. The Screwpull was demonstrated at the 1979 International Correspondence of Corkscrew Addicts Annual General Meeting in London. All the members gave it a try and were asked to give opinions and to criticize any faults they could find. The only fault that could be considered for improvement was that before the cork was fully extracted it would start to turn, causing the operator to finish the job by pulling it out by the handle with the screw handle. When Allen got that information from the ICCA, he started experimenting with different ideas to solve the problem.

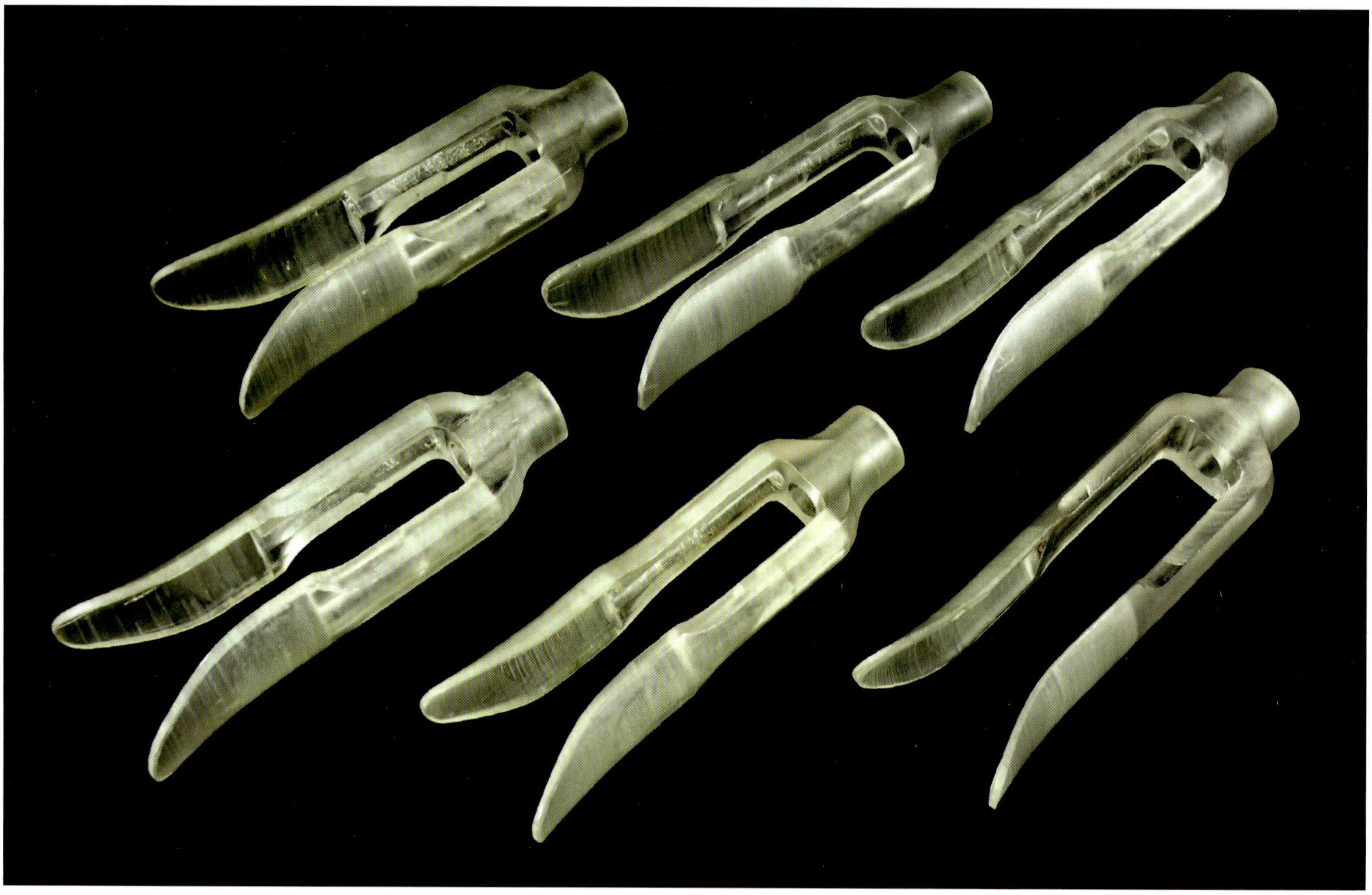

A few examples of some of Allen's experiments for preventing the cork from turning.

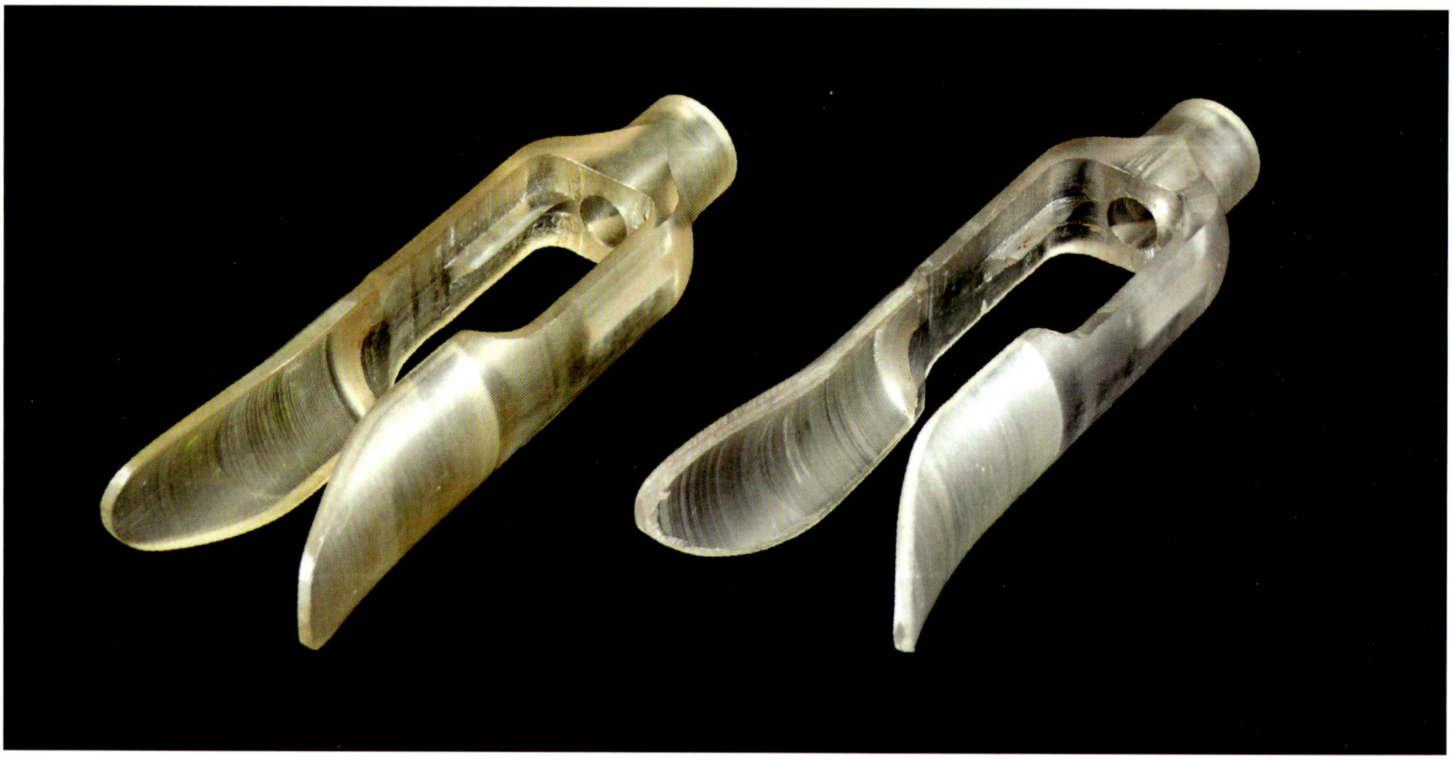

The final design has a ridge on the side of the insides of the frame to put enough pressure on the cork as it rode up the screw to keep it from turning. These two prototypes have the ridges on a Version 1 size frame. The final production model of Version 2, with the ridges, has a slightly smaller frame, as seen in the next photograph, and Version 3, even smaller. The Version 1 Table Model with the ridges was never put into production. Version 2 was the first production model with the ridges. The prototypes shown here, over three decades later, still have cork and wine residue on the ridges and the bottom side of the top of the frame from their test runs.

Version 2 Screwpull prototype in clear plastic material, c. 1980.

Version 3 Screwpull prototype in clear plastic material, c. 1980.

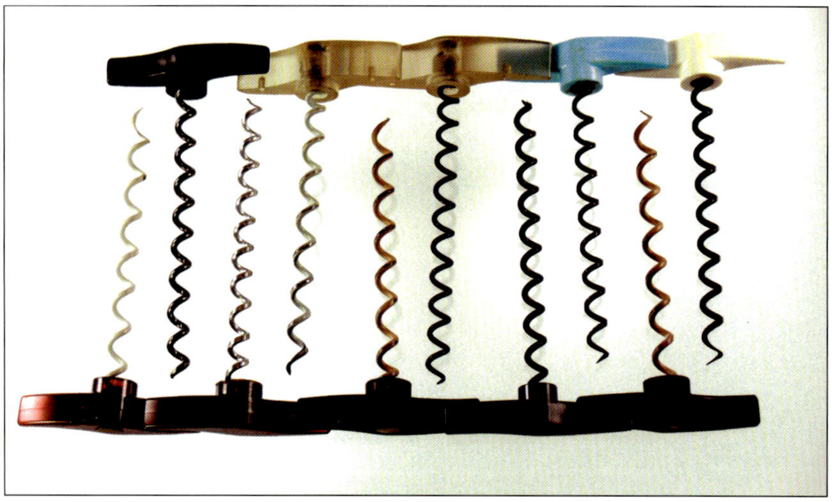

Note the different types of screws Allen tried during his experimental period.

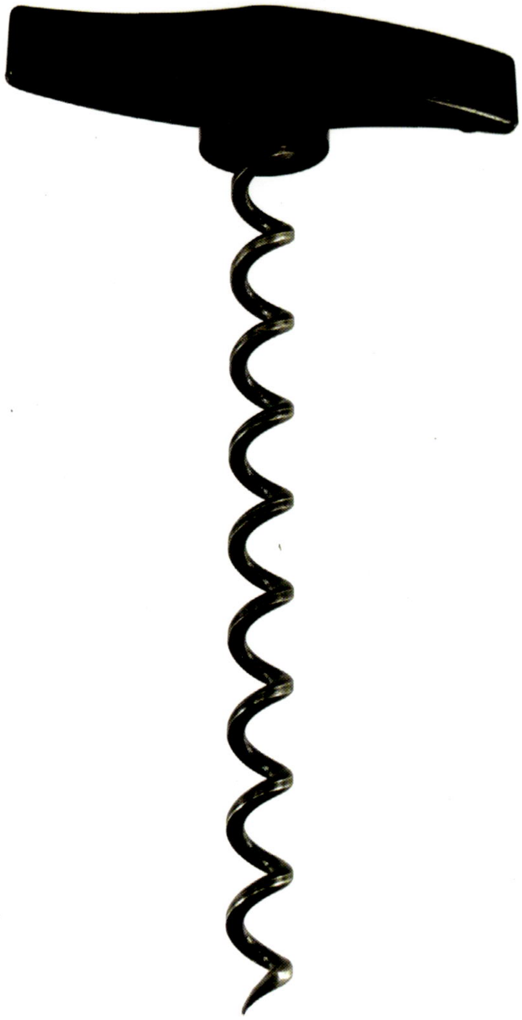

He even experimented with a fluted wire helix.

A few of the first production models had a pitch of 16 mm with a white coating, but from then on the production models had a pitch of 13 mm with a dark brown coating. In December of 1980, Allen had a testing laboratory run a Bezier Test on a typical Screwpull screw. The Bezier Test screw has a pitch of 13 mm with a grey coating. The Bezier Test screw was driven through and withdrawn from more than 2,000 corks and finished in good condition. Some of the coating of the lower half of the screw had worn off enough to expose the shiny surface of the stainless steel, but it was determined that the screw still had its reduced friction capabilities. The Teflon coating process not only coats the surface of the worm but also imbeds the material into the steel, maintaining its friction, reducing qualities even after much of the surface coating is worn off.

Most of Allen's problems were solved with his trial and error methods. Centering the screw in the cork where it would run true through the center of the cork, and allowing the cork to ride straight up onto the screw in a line parallel with the longitudinal center of the screw was an important factor in his design idea. Centering the screw on the cork was a basically simple solution consisting of a bushing or guide in the neck of the frame. The bushing in the final production models was a smooth hole in the neck of the frame the exact same size of the outside diameter of the screw, forming a snug fit and keeping the screw in perfect alignment. The early prototypes had separate bushings made of different materials, such as brass and harder plastics.

Evidenced by his prototypes and March 22, 1983 patent, Allen's desire for "perfect" centering took him on a tributary of experiments of guides. Most of these experiments utilized an "upper" body section, where the screw was not removable and had tracks or rods, ridges or other devices between the upper and lower unit to run the screw straight and center into the cork. These were complex devices and much unlike Allen's thinking. He apparently experimented with these "centering" devices before he realized that the screw point was his problem, not his simple hole-in-the-neck bushing. Once the point was resolved, the simple Screwpull, as we know it today, evolved.

A Screwpull screw after a Bezier Test of more than 2,000 cork insertions and withdrawals. Taped to the top of the handle is a note that reads, "After over 2,000 corks, Bezier Test."

A Screwpull experiment, with a sliding centering device, three-legged flat bottom frame with an aluminum guide fixed to the screw, and the aluminum handle.

The real problem was that after the screw started "running" through the cork, it would travel out of the cork and to the side of the bottleneck or any which way, resulting in a chewed-up cork. Herb then discovered that a differently shaped point on the screw would take the point in a different direction through the cork. Grinding every shaped point imaginable became an obsession until the problem was solved, and solved. Grinding a slow-pitched point on the top side of the wire allowed the screw to pull up and true, rather than down and/or out.

A Screwpull experiment with a sliding centering device, two-legged flat bottom frame with an aluminum guide fixed to the screw, and the plastic fixed handle.

A Screwpull experiment with a sliding centering device, Version 1 frame and handle with a rustless steel, 4-post guide fixed to the screw, and the plastic handle.

A Screwpull experiment with a sliding centering device, Version 1 frame and handle with a brass, 2-bladed guide fixed to the screw, and the plastic handle.

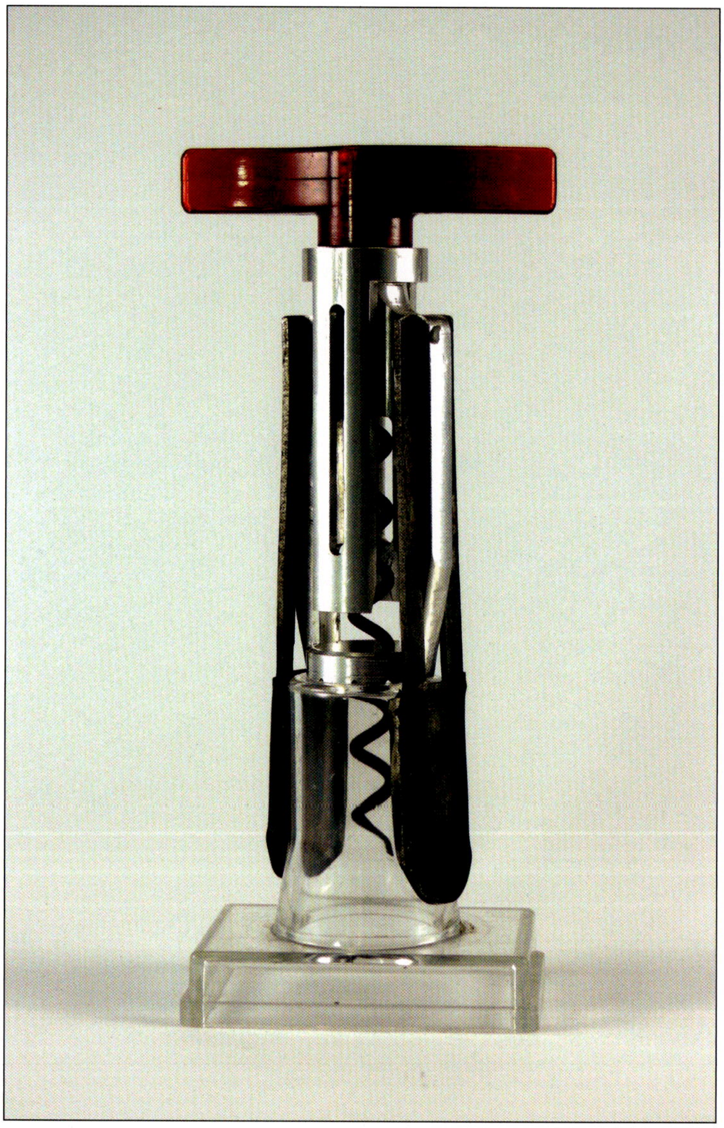

A Screwpull experiment with a sliding centering device, all-metal frame, and guide system fixed to the screw and plastic handle.

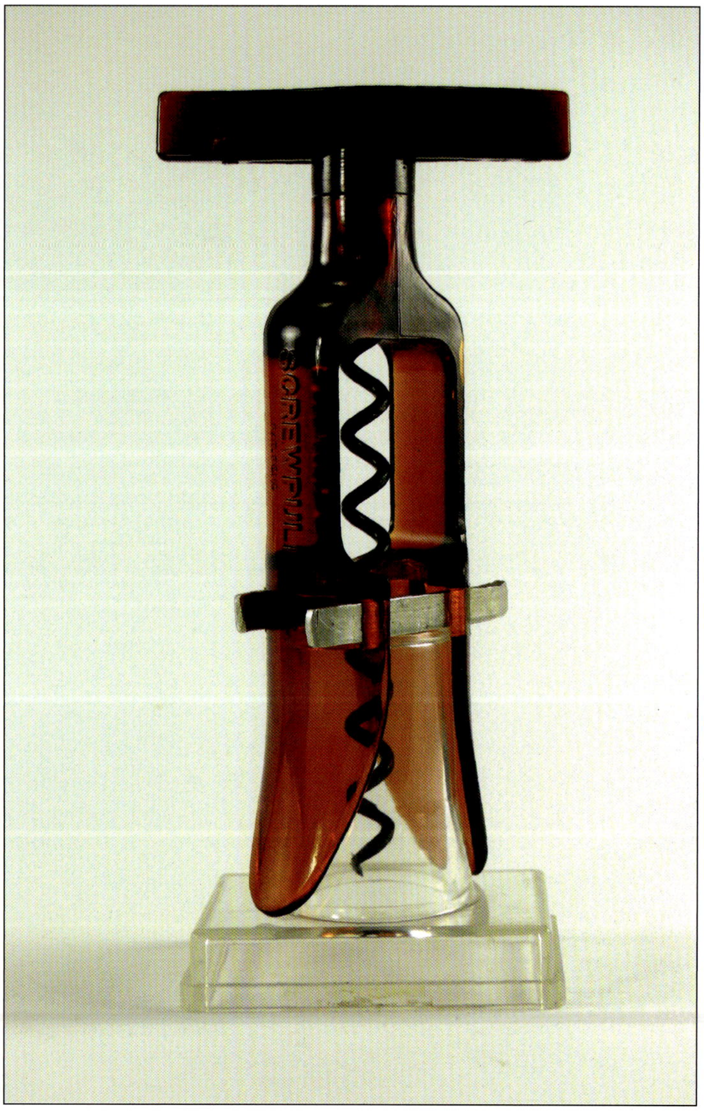

An Amber Table Model Version 1 with experimental bars through legs of the frame designed to not let the legs go out too far and break. This is possible if you do not hold the legs tightly to the ledges on the top of the bottle while pulling the cork. The legs will spread and push the ledges over the top of the bottle and break one leg. This experiment was never manufactured.

Two experimental Version 1 Table Model Screwpulls with devices to prevent the cork from turning prior to complete extraction. The one on the left has two spikes protruding from the underside of the top of the frame and the one on the right has two small curved plastic plates in the same area to try and accomplish the same thing. These experiments were prior to the ridges in the sides of the frame, which became the final solution and was included in the 1983 Patent #4,377,096.

The downwardly facing surfaces of the screw form a continuous helix of constant lead or path ("Summary of the Invention, U.S. Patent No. 4,291,597," September 29, 1981).

High tech engineering and a desire for quality are also evidenced in the frame and handle of the Screwpull by the use of D. E. Polycarbonate Plastic, otherwise known as Lexan. The flexible legs of the frame allow for a tight "grip" on the top of varying-sized bottle tops. Attention was given to every detail of the frame and handle. The relevant patents and prototypes included free-standing, three-legged versions, two-legged versions with flat bottoms that would stand up without a stand, different types of slots in the frames for added flexibility, etc. The three-legged versions made it awkward to remove the cork from the corkscrew. One's fingers had to oppose each other at a 120 degree angle instead of at 180 degrees, as in the two-legged or two-finger port devices. This was before Allen developed the catch ridges that would prevent the cork from turning when you would unwind the screw from the cork. You had to use your fingers in the early versions of the production models to grip the cork while the screw was spiraled out.

The straight, flat bottoms did not grip the bottle well and did not have a comfortable feel. The tapered-flared bottle grips were the perfect shape, functionally and aesthetically, but did not allow the corkscrew to stand up by itself. Instead of compromising the design, Allen developed a stand, similar to a bottle top, for the corkscrew to rest on gracefully.

Screwpull Fact: The Screwpull maintains a place in the New York Museum of Modern Art for its excellence in design.

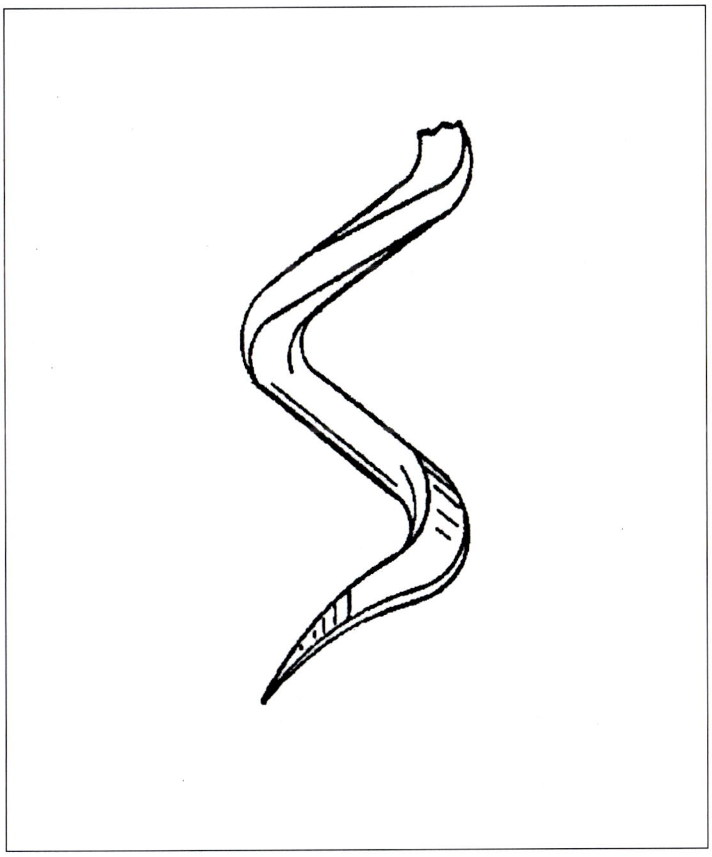

A patent drawing showing the business end of the screw.

A Screwpull Table Model Prototype with a two-legged frame with flat bottoms, allowing it to stand alone without a stand, c. 1979.

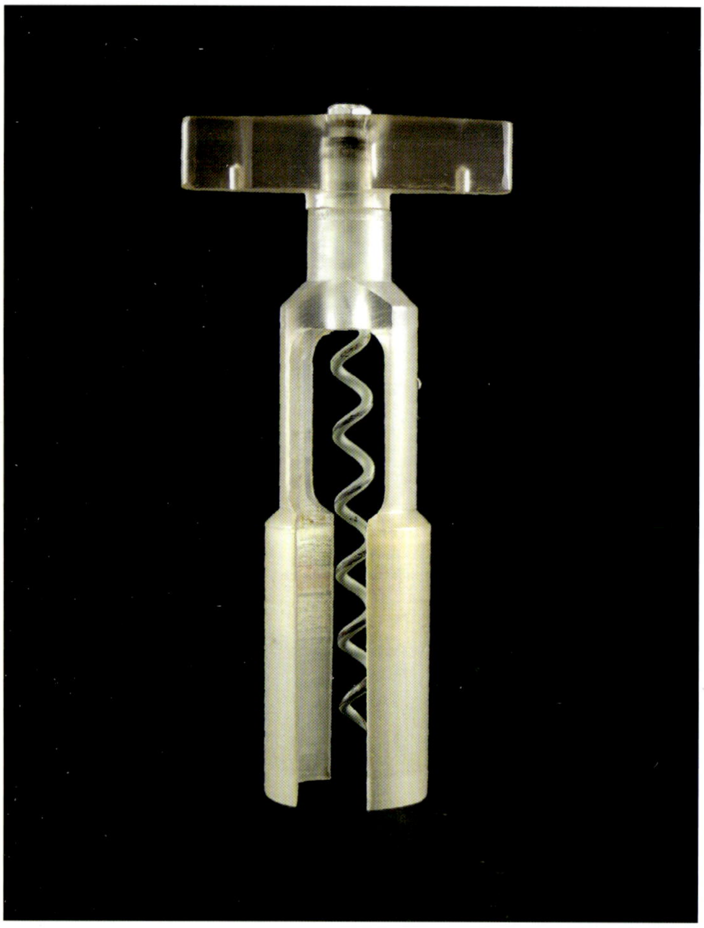

A Screwpull Table Model Prototype with a two-legged frame with flat bottoms, allowing it to stand alone without a stand, c. 1979.

A Screwpull Table Model Prototype with a three-legged frame with flat bottoms, allowing it to stand alone without a stand, c. 1979.

A Screwpull Table Model Prototype with a three-legged frame with flat bottoms, allowing it to stand alone without a stand, c. 1979.

A Screwpull Table Model Prototype with a two-legged frame with flat bottoms, allowing it to stand alone without a stand, c. 1979. The handle is missing.

An Experimental Prototype with stand-alone, flat bottoms and a built-in foil cutter with slot for a knife and cap lifter. Possibly designed for use with a Pocket Model, this idea was included in one of his patents but was never manufactured. Note the black coated fluted worm.

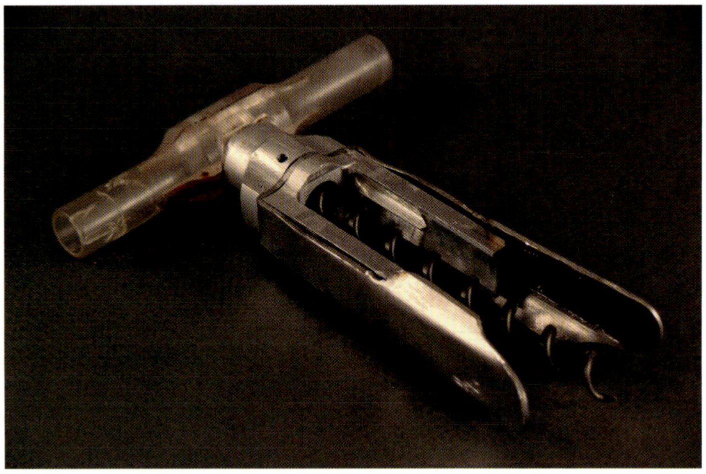

A Screwpull Pocket Model Prototype, all-metal frame with plastic handle/sheath.

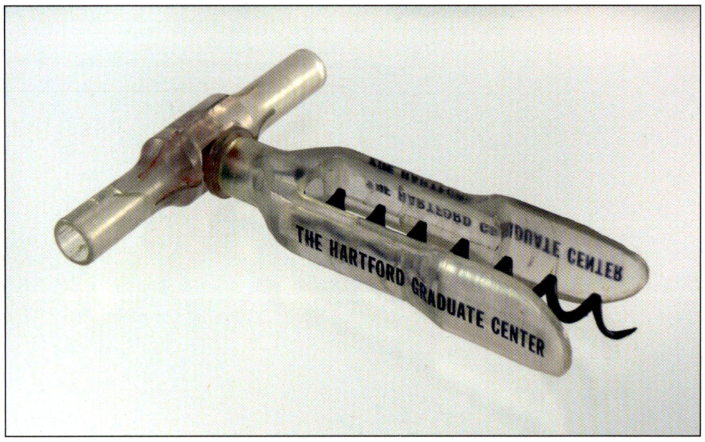

A Screwpull Pocket Model Prototype, clear plastic frame with plastic handle/sheath. Allen made this prototype for a pocket model to give to Dr. Homer Babbidge, president of the Hartford Graduate Center in Hartford, Connecticut. Dr. Babbidge (All-Right) was the Right of the ICCA at the time. I never did see the final product.

A Screwpull Pocket Model Prototype with a rod coming down from the top forming a half circle where the bottle-top stop ridges would be and going back up to the top, on each side of the frame. The two half circles would form the bottle-top stop and the two parallel rods on each side would keep the cork from turning, hopefully. This experiment was never manufactured.

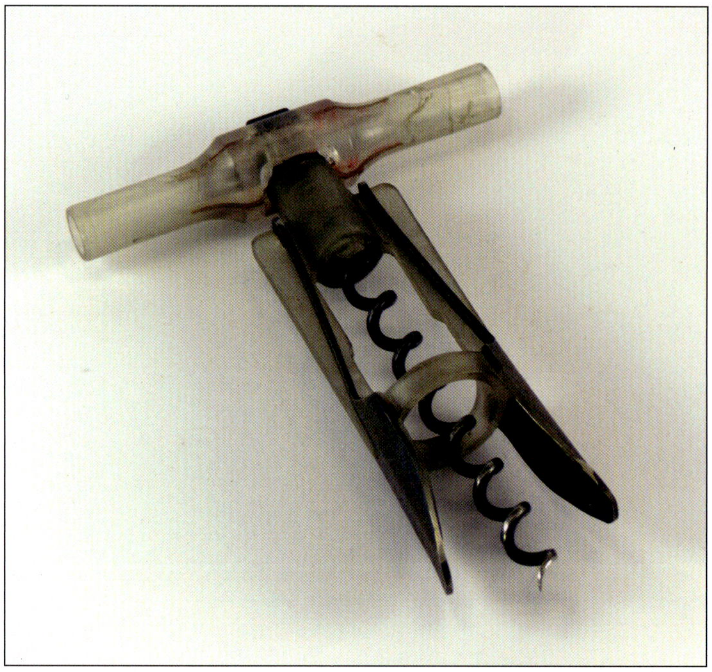

A Screwpull Pocket Model Prototype, all-metal frame with plastic handle/sheath.

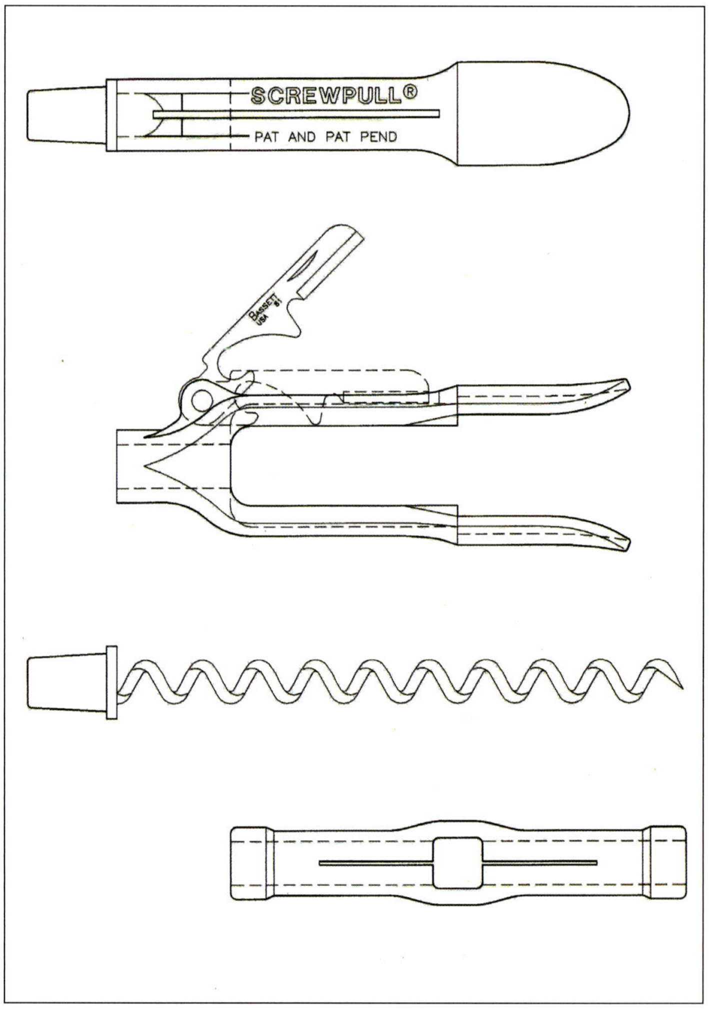

Drawings of Version 1 Pocket Model Screwpull. *Courtesy of DM.*

Experimental Pocket Model handles.

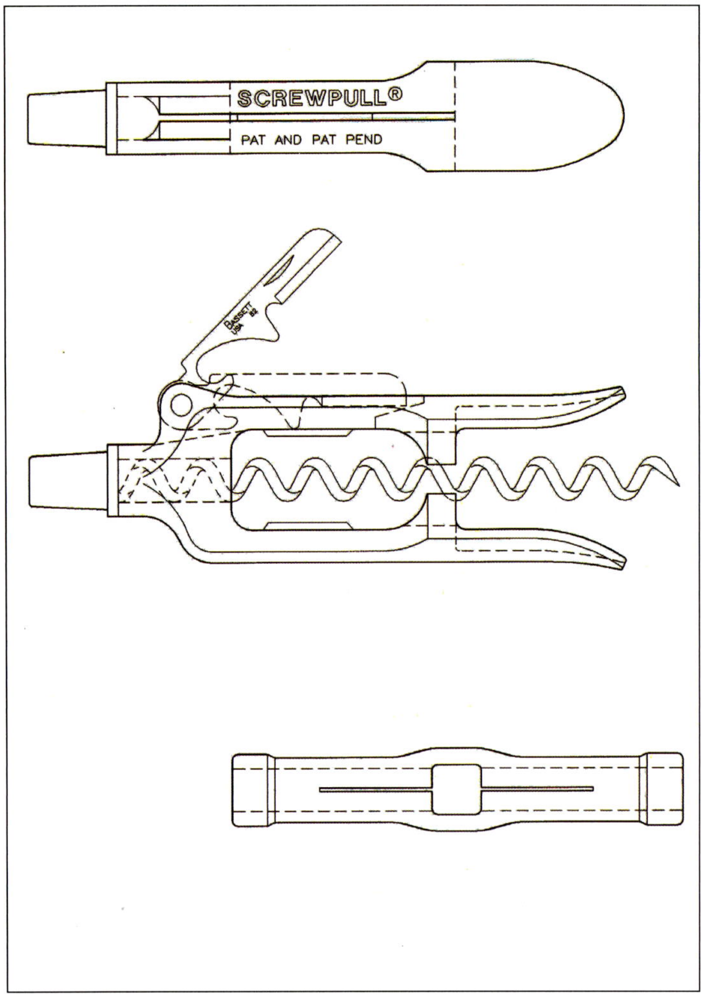

Drawings of Version 2 Pocket Model Screwpull. *Courtesy of DM.*

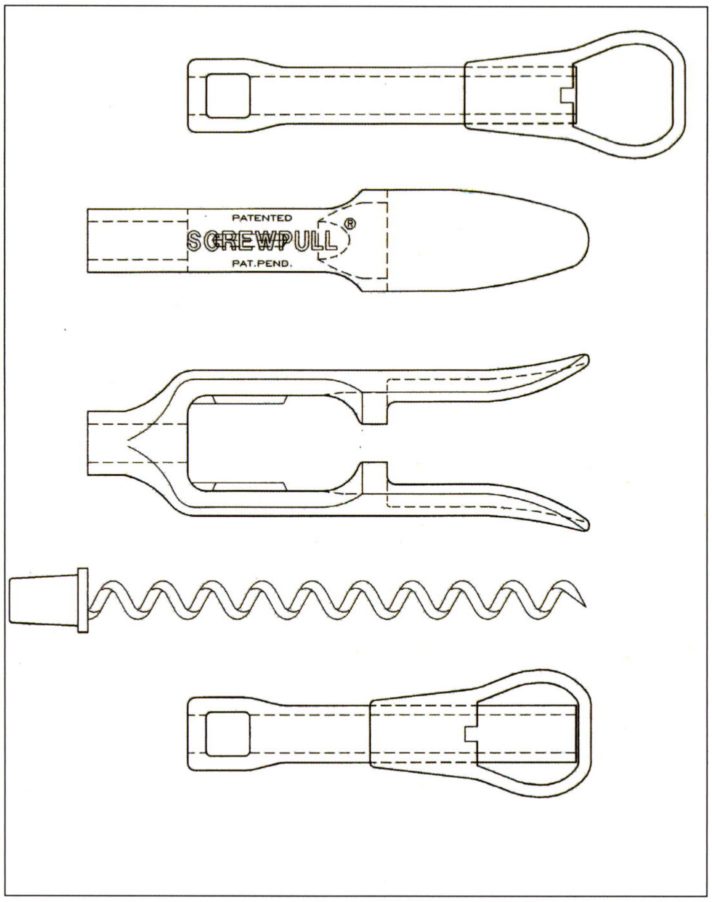

A drawing of Spinhandle Pocket Model Screwpull. *Courtesy of DM.*

The First Spinhandle Prototype.

The photo to the right is a Prototype of the bushing used for some of Allen's early experiments and is shown in the patent drawings. The actual bushing or nut would be cut to length from a milled piece such as this. The groove spiraling down the "bolt" is where the coated screw would run through the nut. Therefore, a short piece of this milled bolt would be screwed into the threaded bushing and become the nut for the screw.

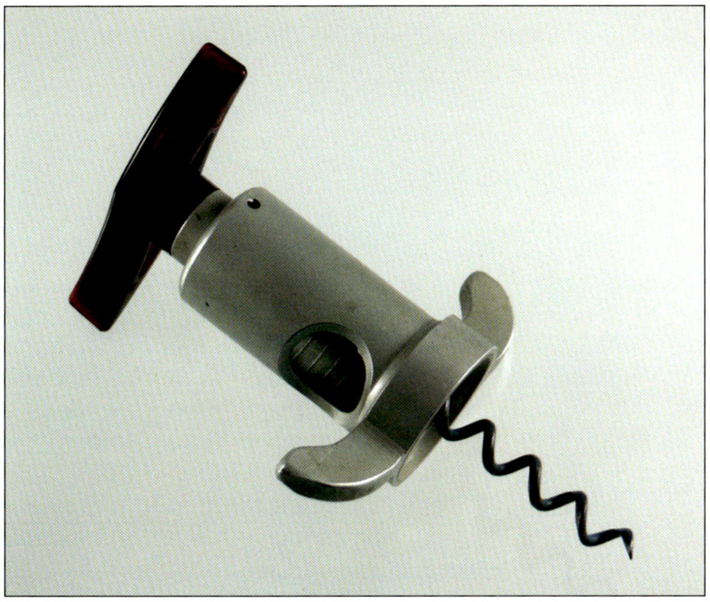

An early prototype machined out of aluminum.

It would be hard to guess what he was thinking with this one. The screw was his idea from the beginning and then he went through many other ideas of what kind of frame he would use to make the screw work to its best advantage. This was an early thought. The screw is not loose in the bushing. The bushing is actually a nut through which the screw winds down. The bushing screws into the top of the aluminum frame and is actually shown on sheets 3 & 4 of the drawings of patent No. 4,276,789. You insert the screw into the cork by turning the handle and continue by turning the frame with its ears until the cork reaches the top of the aluminum frame when it should be fully withdrawn. Since the screw does not turn freely in the bushing, like the final Screwpulls do, it runs up and down like a bolt in a nut. Reversing the action will fully remove the screw from the cork as the nut is pushing the cork away from the frame as you turn the handle in reverse. Allen spent quite some time with the nut idea, as evidenced by his prototype examples and getting a patent for it, but then decided it would be too complex in the manufacturing process and surely not as people-proof as his final solution.

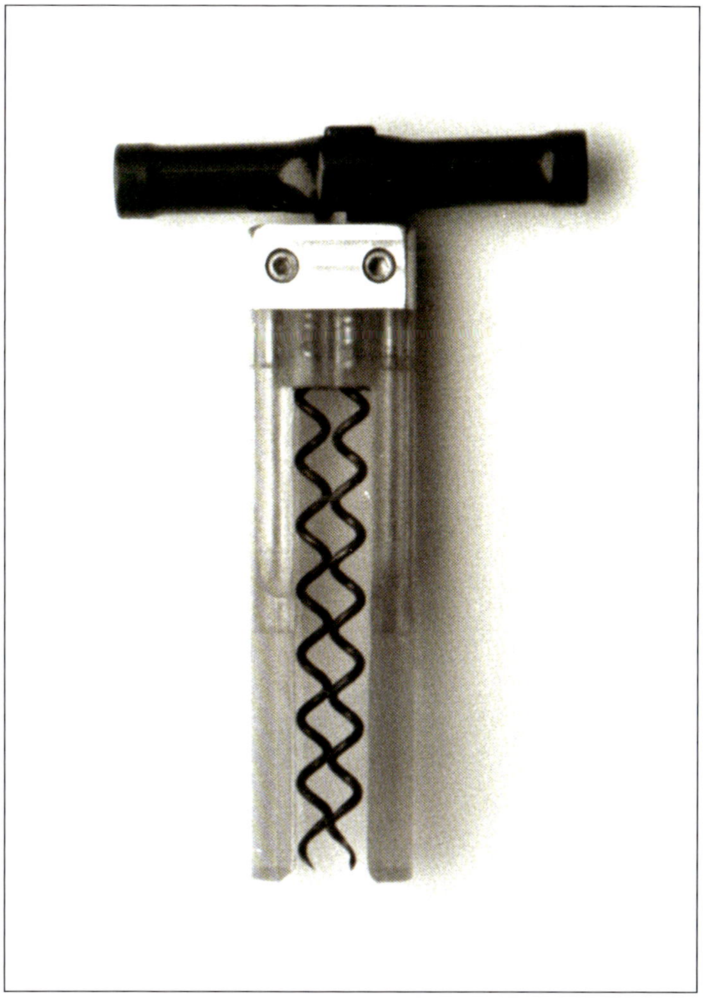

A two-screw Screwpull prototype.

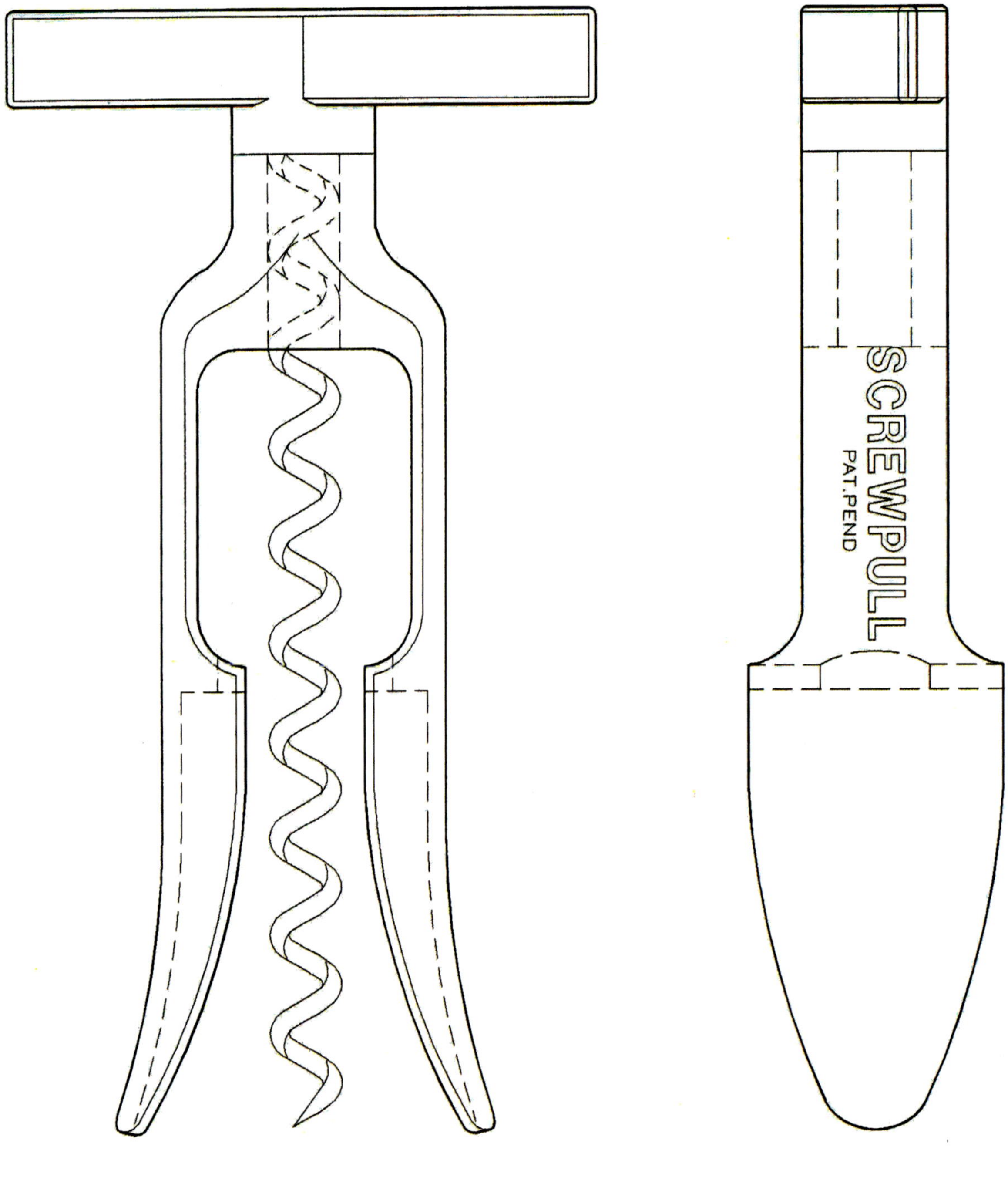

A drawing of a Version 1 Table Model Production Model. *Courtesy of DM.*

Chapter 3: The Screwpull Production Models

There were three apparent changes and improvements to the frame section of the Screwpull after the original production model:

1. The lower tapered area of the "legs," of the frame are 1 3/8" wide.
2. The width of the lower tapered grip area of the "legs" was reduced from 1 3/8" to 1 1/8". For what reason I am not sure, other than the legs just did not need to be the larger size to perform their designed function.Vertical ridges were added to each side of the inside of the upper frame as a "catch" means to resist the rotation of the cork, in both clockwise and counter clockwise directions. On new, not so tight corks, the bottle would release its grip on the cork prior to complete extraction and the cork would start to turn. These ridges would catch the cork and allow the cork to continue its ride to the top for an uninterrupted complete removal. This "catch" means was included in the March 22, 1983 patent, No. 4,377,096.
3. The upper body thickness was reduced from 3/4" to 5/8", and the width of the lower grip area was reduced to 1".

Production Models of Version 1 Table Model. The earliest version on the right with a silver screw instead of the later version with brown coated screw, and the raised logo is not "painted."

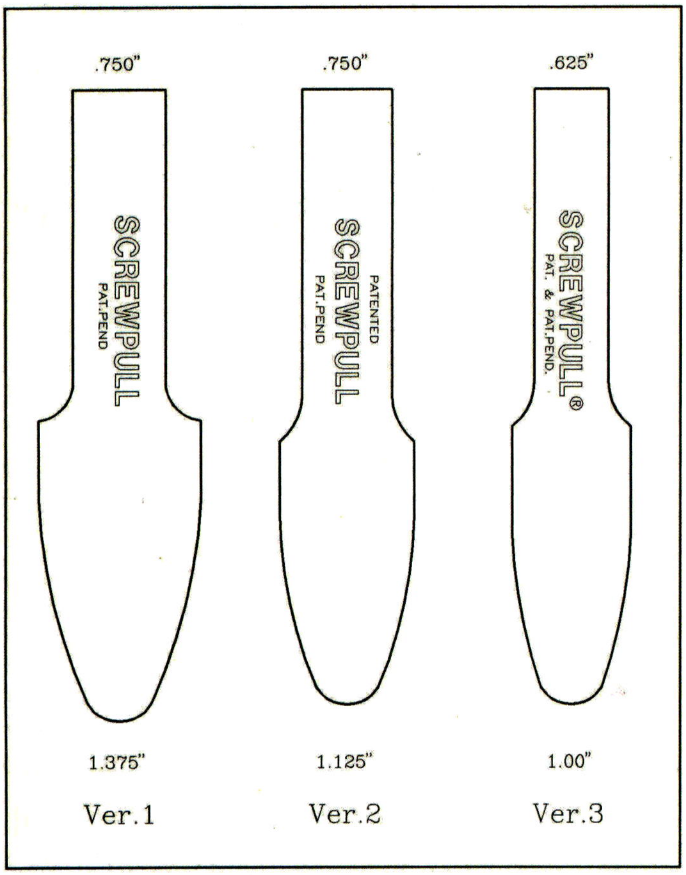

A drawing of the three versions of Table Models with the major production changes. *Courtesy of DM.*

Production Models of Version 2 Table Model. The earliest version on the left has the ridge that prevents the cork from turning, extending to the top of the frame.

The Classic Table Model, model numbers were 1010 Amber, 1020 Burgundy (A Transparent Red), 1030 Black, 1040 White, 1050 Chrome Plated, and 1060 Gold Plated. Just after launch, the pocket models, 2010 and 3010, were only manufactured in black. Later, 2020 and 3020 Red (bright opaque red) were introduced along with 1090 Opaque Red in the Table Model. Model 5010 signified custom printing on any of the above models, and 4010 referred to a display unit for retail sellers' shelves or counters. After 1990, other colors eventually appeared on the Table Models and Pocket Models as well.

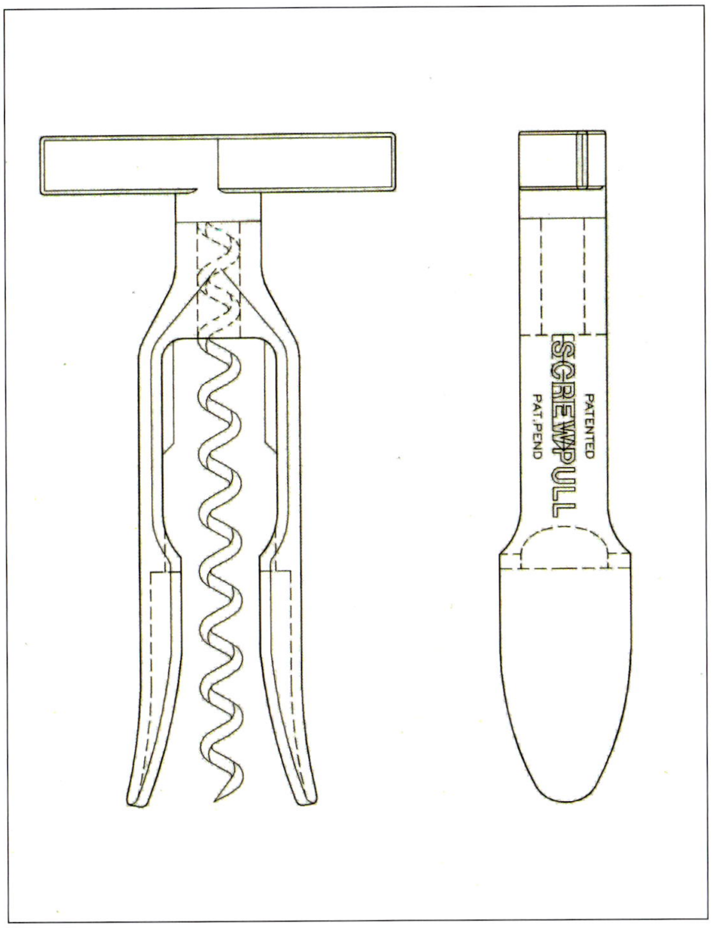

A drawing of Version 2 Table Model Production Model. *Courtesy of DM.*

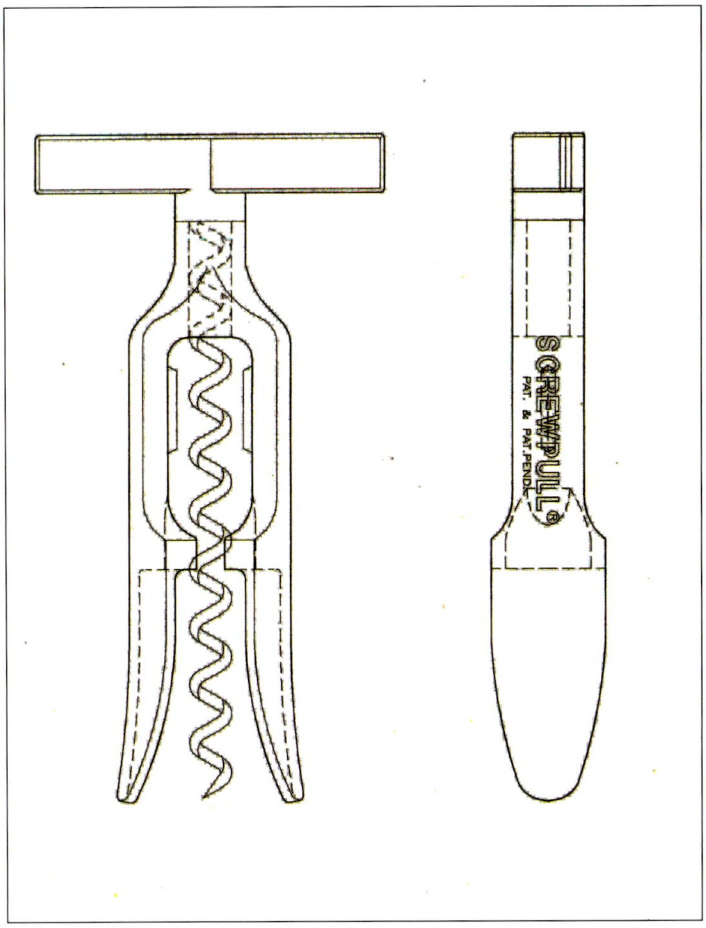

A drawing of Version 3 Table Model Production Model. *Courtesy of DM.*

Production Models of Version 3 Table Model. Note the smaller frame with bottle-top stops extending farther out and the shortened cork catch ridges of the Table Model on the left. This model also became the standard for the Spinhandle Model shown on the right with the only difference being the cork catch ridges extend to the top of the frame, similar to the Version 2 Table Model.

Table Model #1020, Burgundy.

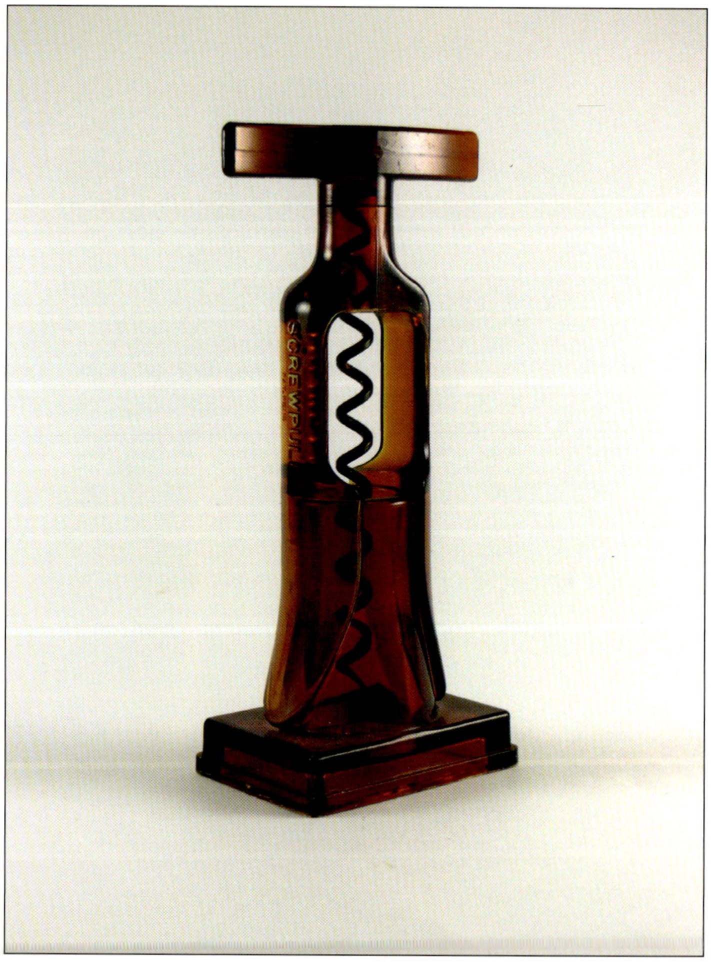

Table Model #1010, Amber (Some of us called the Amber color "Root Beer" but that was never an official color name). Amber and Burgundy were transparent colors, where the other colors were an opaque material.

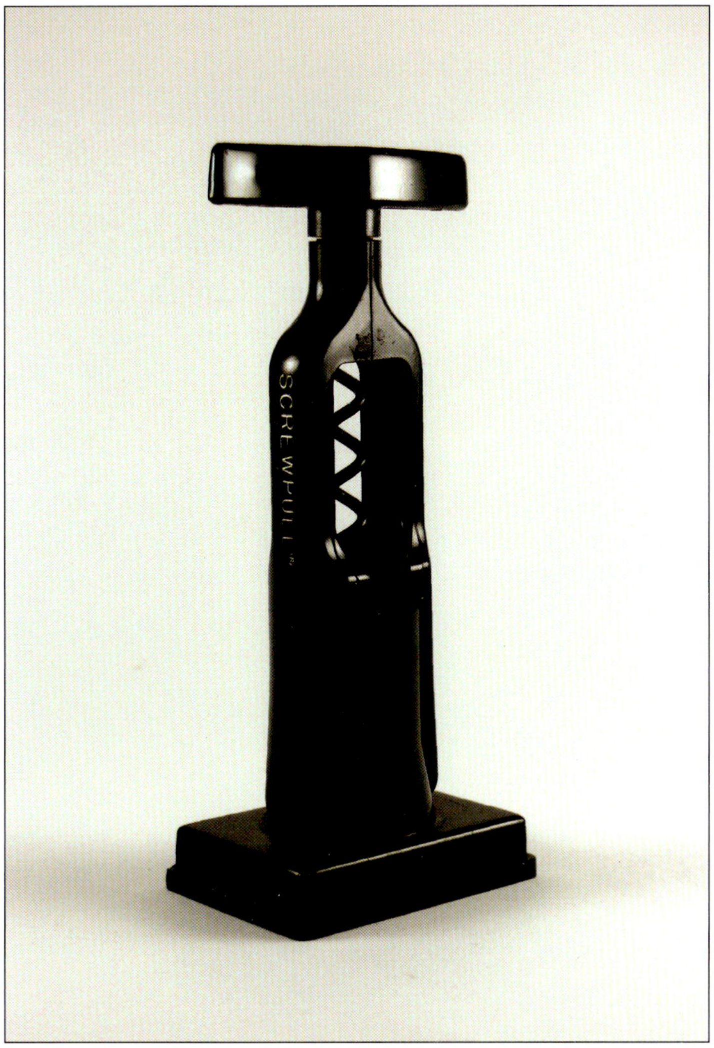

Table Model #1030, Black.

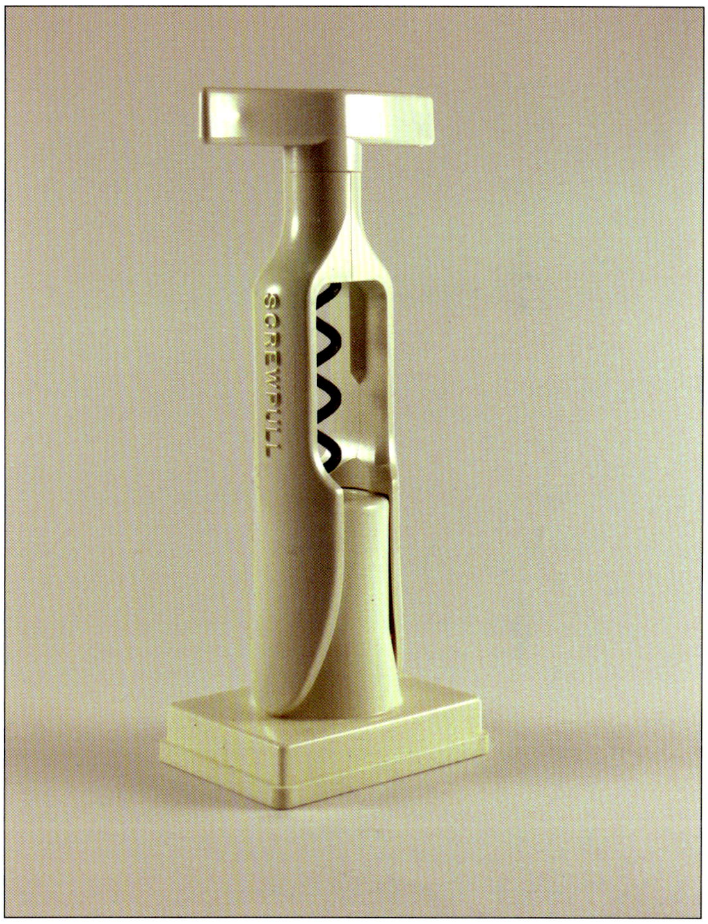

Table Model #1040, White.

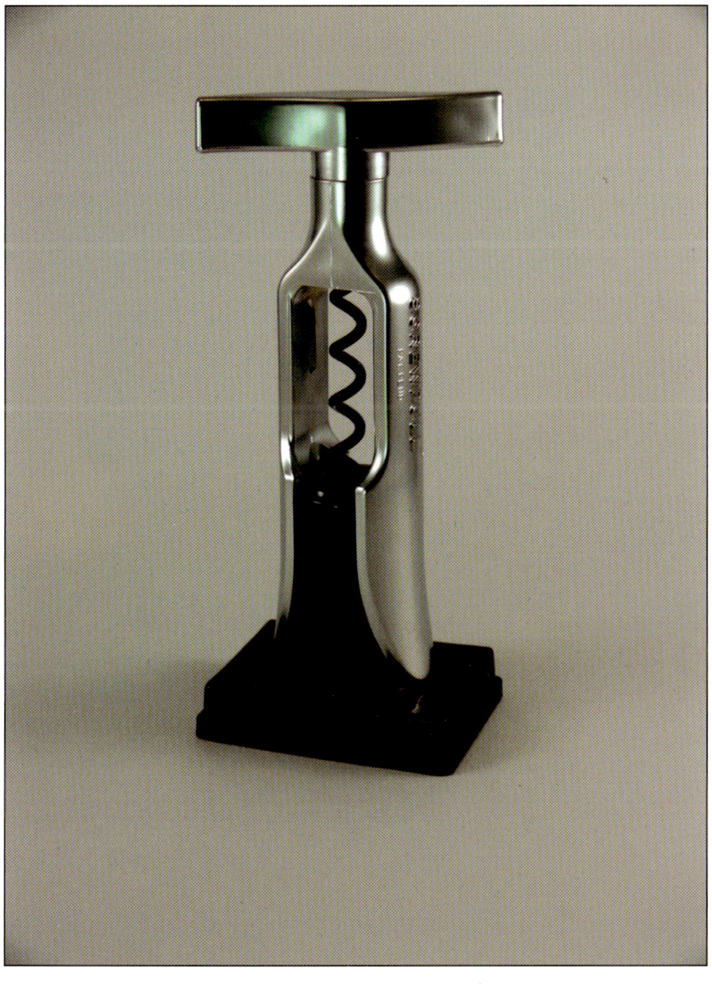

Table Model #1050, Chrome Plated.
Courtesy of Loretta Haren

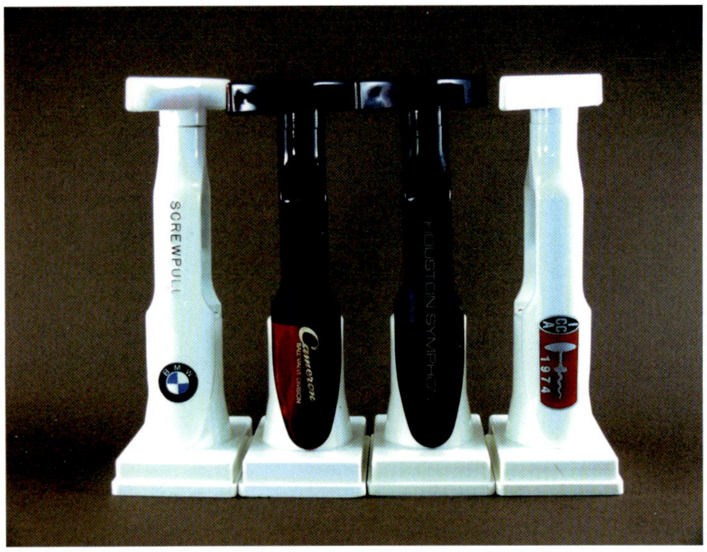

Table Model #5010, Custom Printing.

A dark oak Screwpull was manufactured by a Swiss company with permission from the Hallen Company. A limited edition were made.

The Hallen Company had a company in Switzerland manufacture the Screwpull out of beautiful woods. Allen also gave permission to John Hardy (a jewelry designer in Bali) to make a limited edition of Screwpulls from an exotic wood and silver. The handle and stand are made from Macassar Indonesian ebony hardwood found in the Celebes Islands. Macassar is a hazel-brown black striped wood. The frame is covered in solid silver dots. A very few of these were available from Neiman-Marcus in Houston. An article by James Imbrogno about the John Hardy Screwpull appeared in the April 1997 issue of *Playboy* magazine, giving information on where they could be purchased (see *Playboy* ad on page 174).

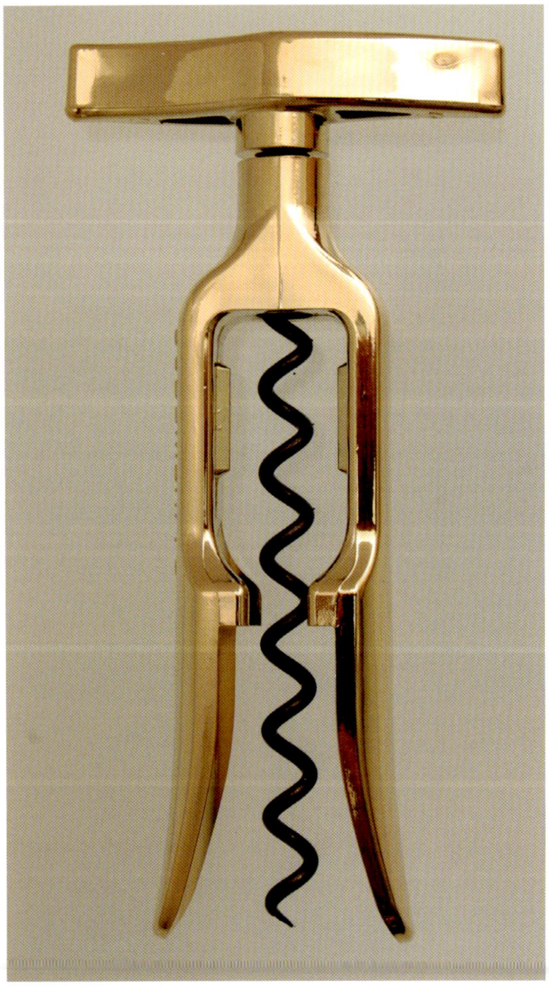

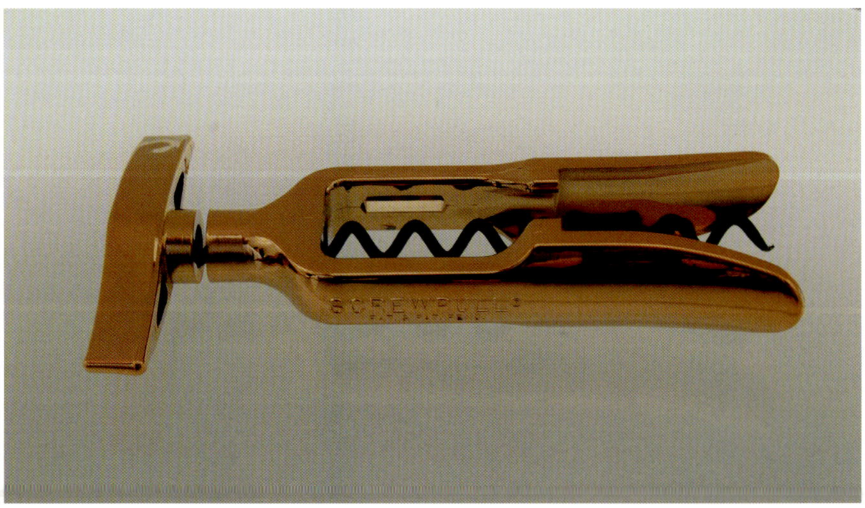

Table Model #1060, Gold Plated. *Courtesy of Mirth-Right Don Bull*

A 2000 Bicentennial Table Model by LeCreuset.

A silver and wood Screwpull crafted by Jeweler John Hardy in Bali, with permission from Herb Allen. This style of Screwpull was produced in a very limited edition from Macassar, a hazel-brown, black-striped wood and sterling silver.

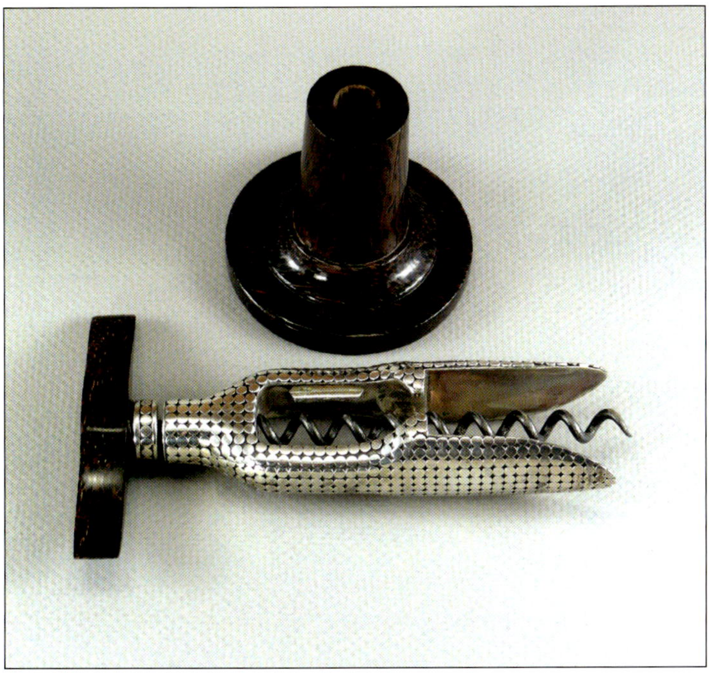

Several different models emerged from the original Classic Table Model. These are the pocket models. The first one to appear had the same size screw as the table model but had a much smaller frame. The removable handle slipped over the screw as a sheath and protected your pocket and leg from the sharp point of the screw. The handle socket was centered and fit onto the square top of the screw to form a T-handle corkscrew. Since the frame is significantly smaller than the table model, the cork "catch" means changes were made to a double ridge on each side to form a channel to receive the cork and the sheath handle, snugly, when not in use. This model went out of production about 1990.

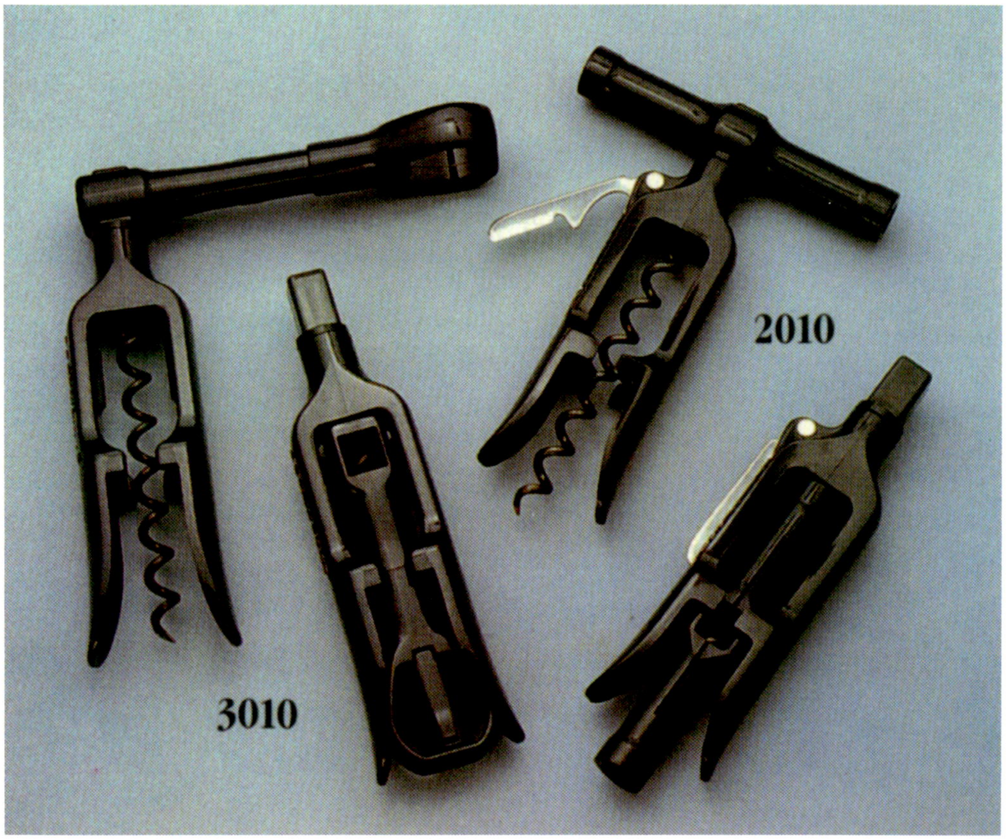

The Pocket Models as shown in a 1981 Hallen Company brochure.

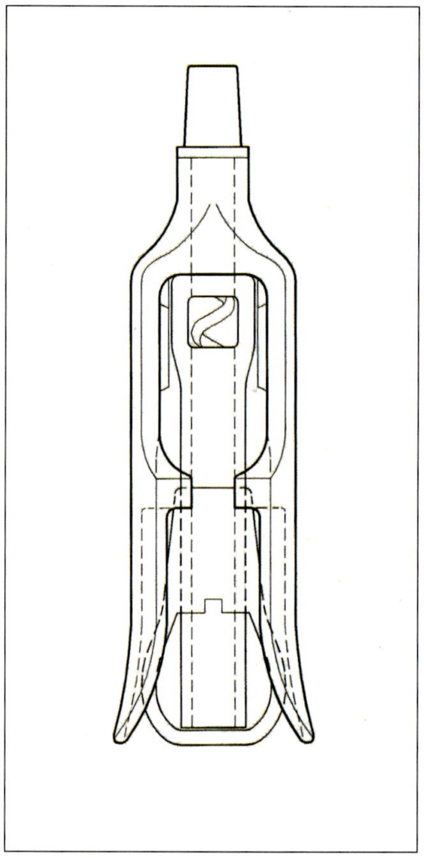

Screwpull Pocket Model drawing. *Courtesy of DM.*

Screwpull Pocket Models.

Shortly after the pocket model went into production, a new pocket model was introduced, the Spinhandle Screwpull. For all practical purposes, the frame was the same size as the table model. The neck and width of the upper frame of the table model is 3/4", where the pocket model measures about 5/8"; however, the thickness of the upper legs of the pocket model is greater than the table model. The offset that resists the top of the bottle is shaped differently to receive the sheath-spinhandle and maintain a snug fit so it does not slip out of its seated position. The handle socket is located at one end of the handle, rather than the center, and has a fingerhole in the other end. The fingerhole pops out an extra inch for greater leverage.

This system allows you to spin the cork out of the bottle with one finger.

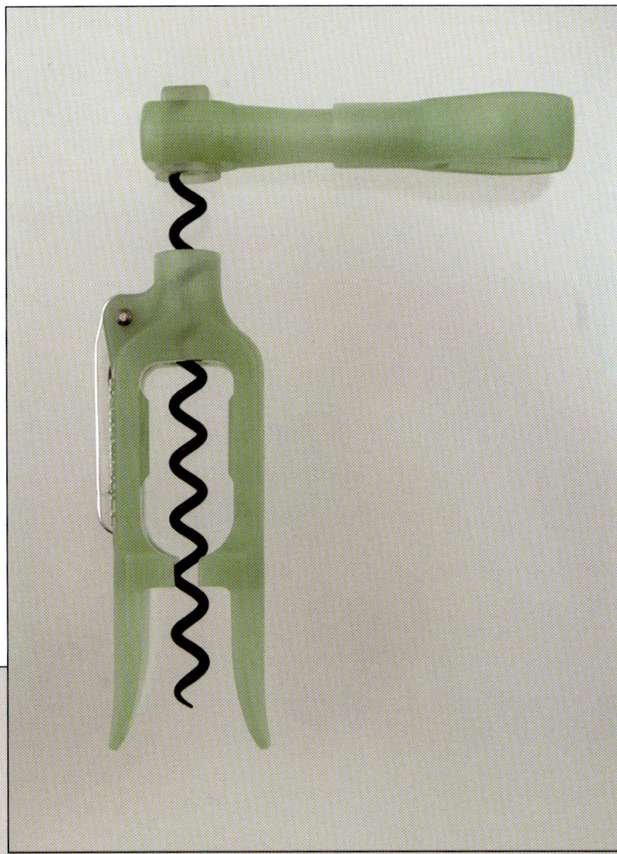

Opaque Light Green Spinhandle Screwpull. *Courtesy of Don Bull.*

Screwpull Pocket Model with instructions. *Courtesy of Don Bull.*

Both versions are equipped with a pull-out, foil cutter knife with a cap-lifter. One early version of the Spinhandle model does not have the knife cap-lifter. The pull-out, foil cutter knife with a cap lifter was added to the production models shortly after the launch of the pocket Screwpull. There were very few made without the knife/cap lifter feature.

All the Table Models, Pocket Models, and Spinhandle Models have in common is the coated screw. The screw is 3/8" in outside diameter and exactly 5" long, from the bottom of the handle at its bearing surface to the tip of the point. The dimensions of the screw have not changed since the beginning of production.

The length of the body frame, from the top of the handle-bearing surface to the tip of the "legs," is exactly 5" on all table models since the beginning of production. The length of the pocket model frames is 4 1/2" on all models except one version of the Spinhandle, which is 5" and is the forerunner of the latest Version 3 table model. The Spinhandle started production with the 4 1/2", Version 2 Pocket Model Frame, but sometime around 1986 it was produced with the Version 3, 5" frame and continued with this frame until LeCreuset launched the Triology Version of the Spinhandle.

An opaque Red Spinhandle Screwpull.

A custom order for the Hugh Johnson Wine Accessories Company was an opaque green Spinhandle Screwpull and a Foil Cutter in a box lined with green cloth.

The Version 2 and Version 3 Spinhandle frames. The 4 1/2" frames are on the left and the 5" frames are on the right.

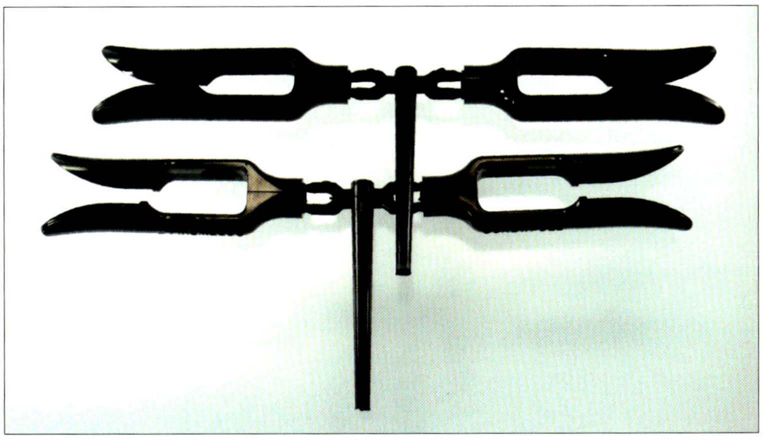

Two completed, two-frame, Version 2 Spinhandle frames out of the injection mold before being separated from the vents.

The Table Models started shipping in 1979, the Pocket Models in 1984, the Spin Handle Models in 1985, and the Foil Cutters in 1986. The Lever Models started shipping in 1987. Lever Models were manufactured from 1984 on, but the major shipping records started in 1987.

From 1979 to December of 1988, there were more than one million corkscrews shipped to the US alone. These included the Table Models, Pocket Models, and the Spinhandle Models. In the year of July 1, 1986 to June 30, 1987, 150,000 Foil Cutters were shipped. From July 1987 to December 1987, the Hallen Company shipped 4,000 Lever Models. These numbers are from the early years of the Hallen Company records and give you an idea of their volume at that time.

Today, as estates are being dispersed, many of these earlier models are being recycled through eBay and other websites.

Chic Monsieur. *Courtesy of L'Atelier du Vin.*

Chapter 4: Screwpull Knockoffs

The Screwpull was such a success, from the time if its launching into the market in 1979, that producers from all over the world have been imitating it ever since. In 1985, Herb Allen had 18 knockoff patent infringement cases in the courts at one time, and just a short time later new versions were coming into the market with increasing regularity. In 1983, more than 30 versions came from Holland, Germany, Japan, Hong Kong, Switzerland, Italy, Brazil, France, Denmark, and the USA. Since the patent ran out in 2001, there are hundreds of different designs of these copies now on the market.

The struggle with patent infringement was both sweet and sour for Mr. Allen. He truly enjoyed the "fight" and had the means to stay with it, but the imitations were spoiling many of his markets, particularly foreign markets. The Dutch "Brabantia" and the German "Sieger" were particularly eating into his European market.

SIEGER + TELESKOP, Germany

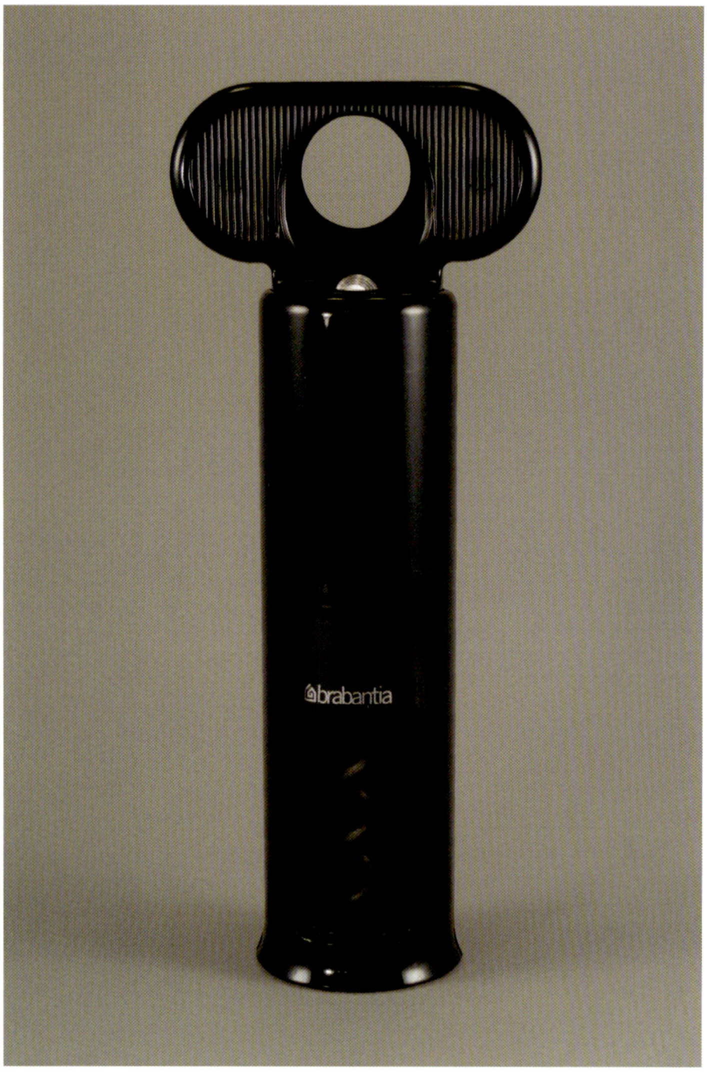

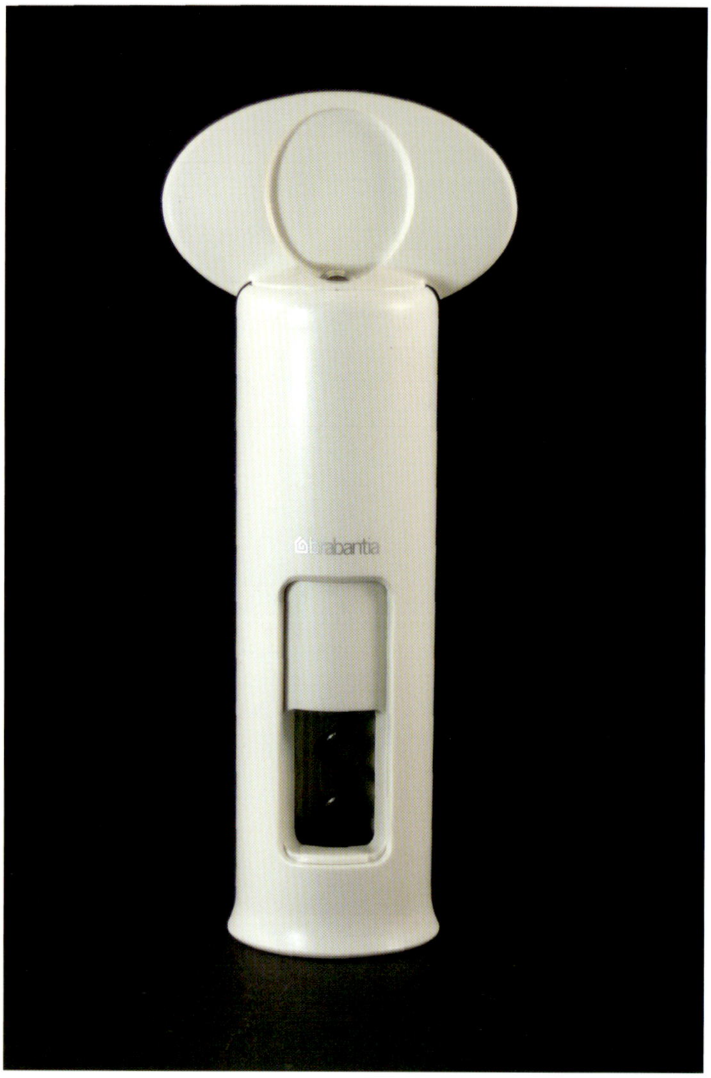

The Brabantia, Holland. This is the one that gave Allen so much trouble with patent infringement.

Several other versions of the Brabantia that came later. The one in the center is called the Obelisque, the one on the right is called The Napolean.

Japanese knockoffs were also a real nuisance, as they were not manufactured in any one place under a name, with the exception of the "Dansk." Each part was made by different people in different locations, assembled in another location, and then exported to countries such as Canada with nothing more than the importer's name stamped on the box and a sticker on the corkscrew reading "Made in Japan." By the time a trace was made back to the responsible parties, they no longer existed.

Many of the imitations are flawed in design in some manner or another, in an attempt to get around patent infringement problems and/or to manufacture them for a lot less cost, such as using inferior plastics, other types of friction-reducing processes of coating or plating the screw, or giving no attention to the point and its role in the design. Some coatings were nothing more than paint, in lieu of the Teflon-coated screws. Using telescoping or other extraction-assistance devices are examples of differences that occured on variations being marketed. Many knockoff corkscrews were and are selling for much less than the the Screwpulls.

Telescoping designs and spring-loaded, extraction-assistance devices are rather gimmicky devices that tend to be problem-prone and are irrelevant on a self-pulling corkscrew. A small percentage of the knockoffs are high-quality corkscrews, producing all the features of the Screwpull, but at half the cost. Each knockoff may be analyzed for its good and bad features quickly, visually, but it takes several "cork pull" tests to discover the good or weak points.

Differences in the products are worth noting. Make intelligent decisions accordingly, particularly when selecting a corkscrew that you want to be of high quality and lasting duraction.

To be copied or imitated by so many companies just has to be a compliment, but it sure does hurt your market.

–Herb Allen

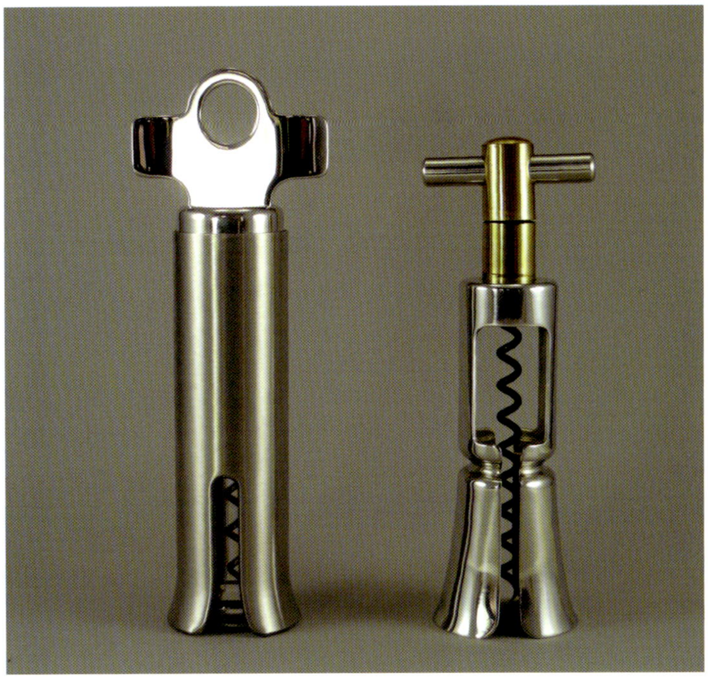

Left to right: The Rosle, China, and the Dansk, Japan.

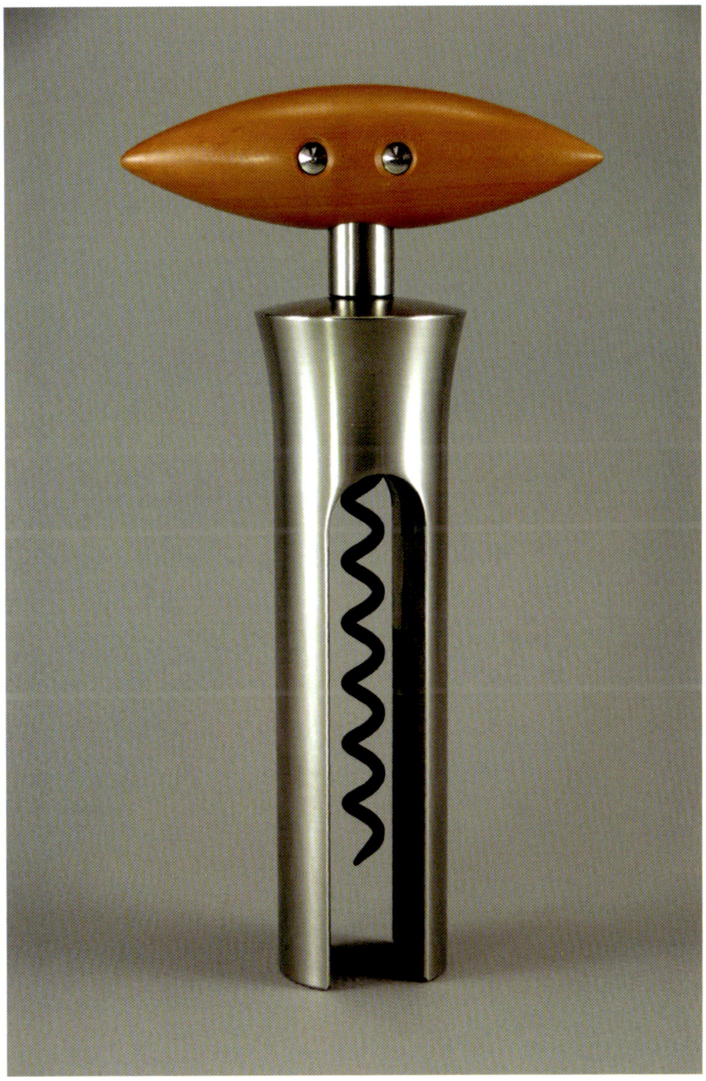

GOLIM - AK7-16-00-01, marked "Akantus----D00," Germany.

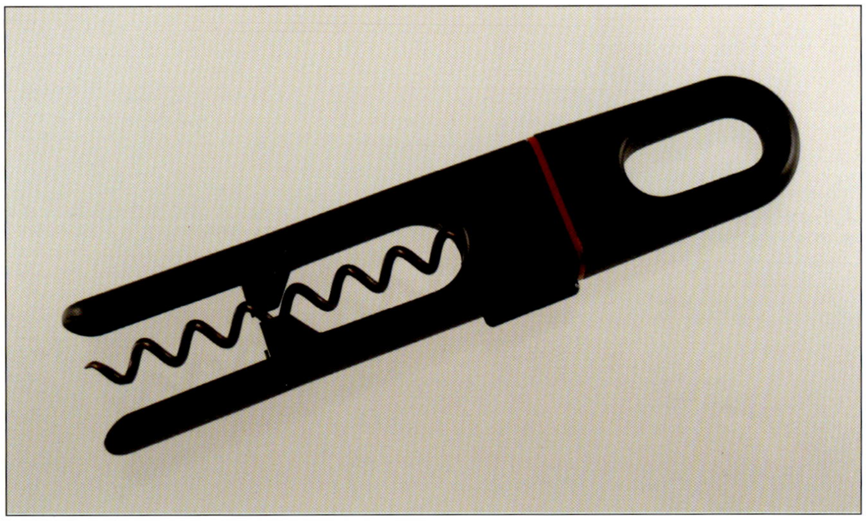

Rosti Mepal "IDEAS," Denmark.

Wenco, Italy.

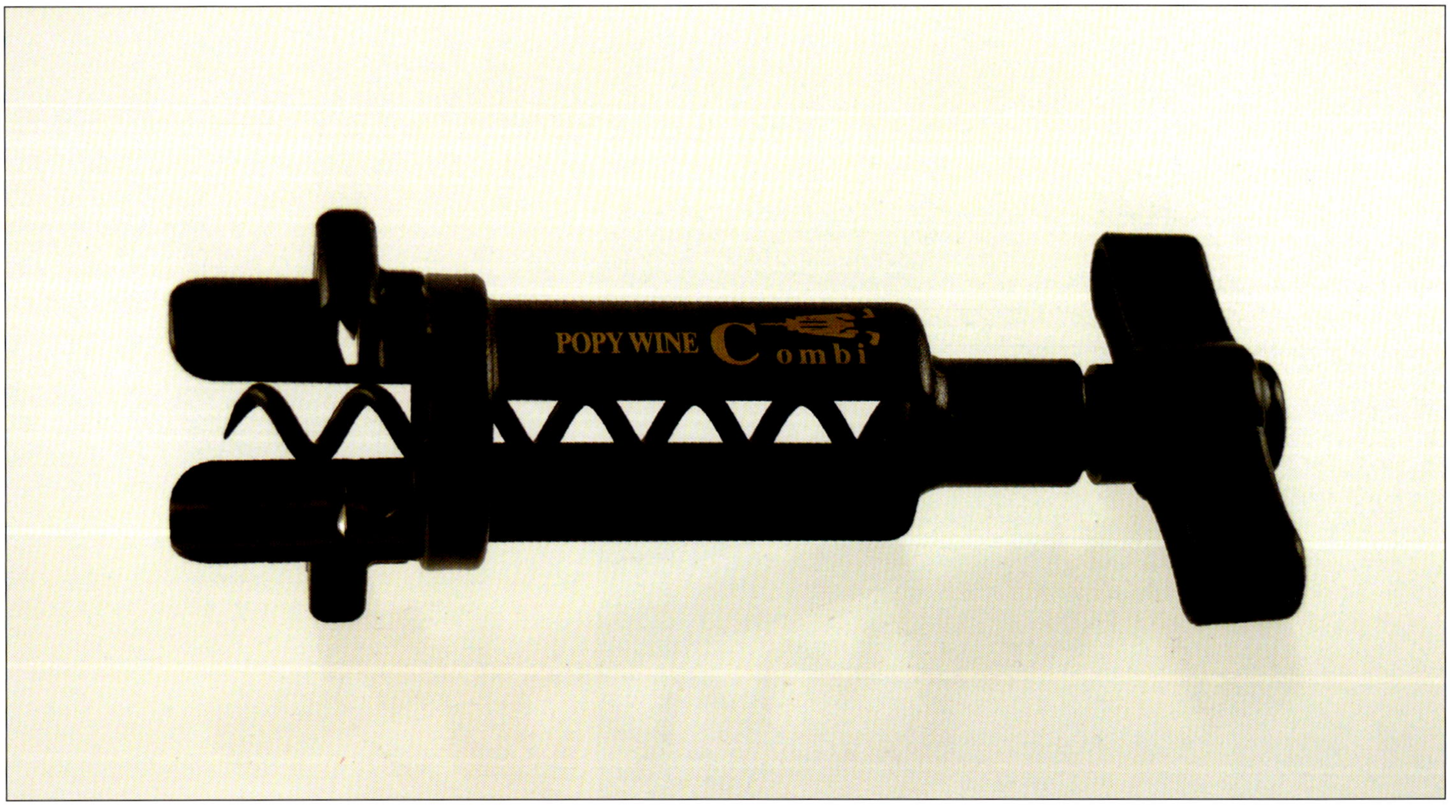

POPY WINE Combi, with foil cutter. A very small self-puller that works well.

Marked "U.S. Patent #B220,855."

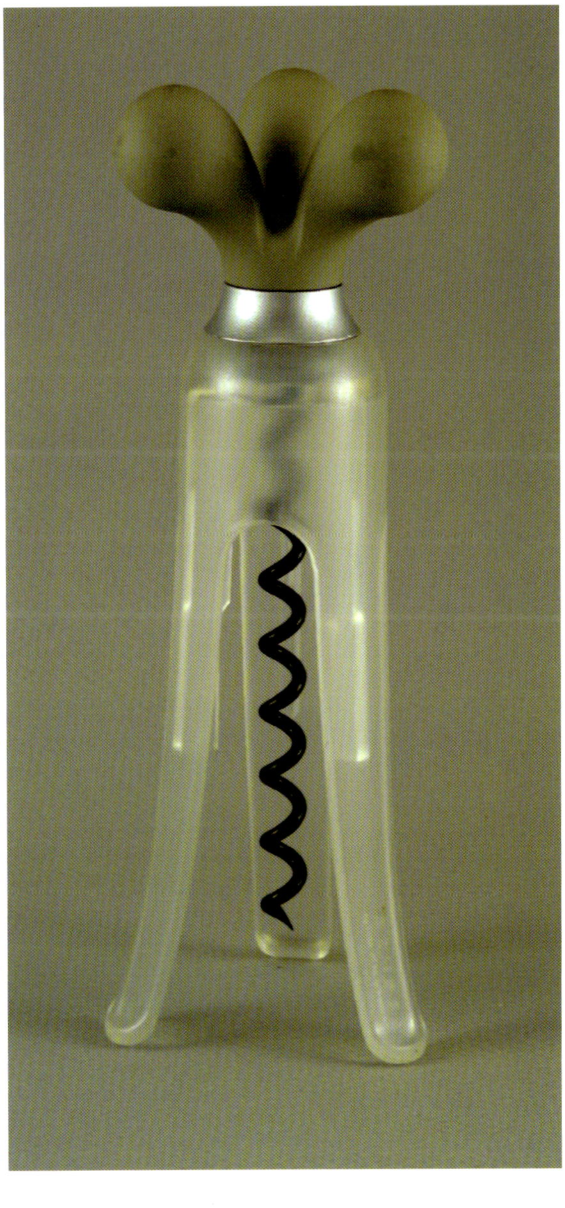

Above and Below:
Marked "Zense," wl7k Design.

A translucent blue plastic WMf.

WMf Korkenzieher, Germany.

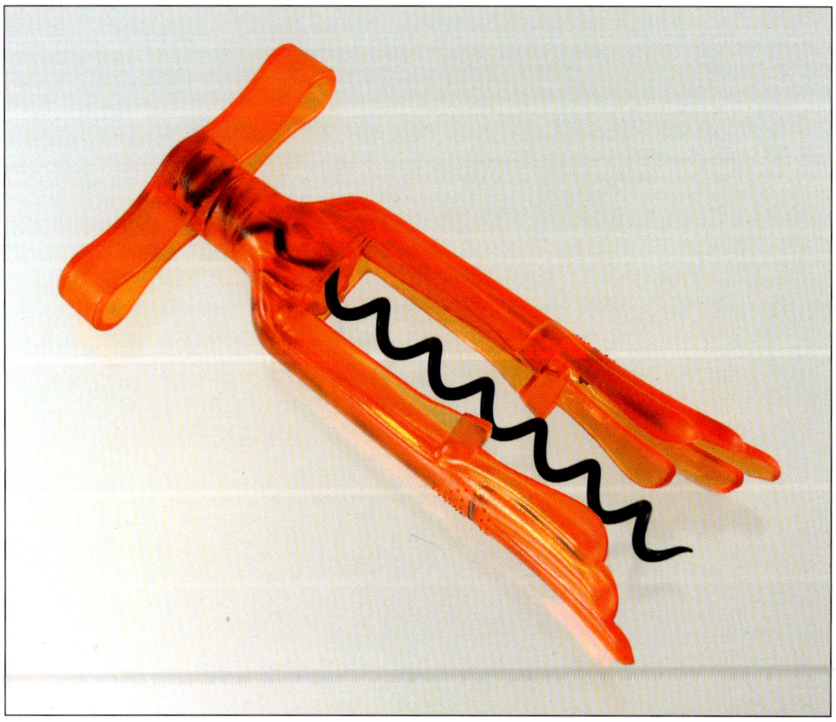

Unknown

Fackelmann, Germany.

WMf, Germany.

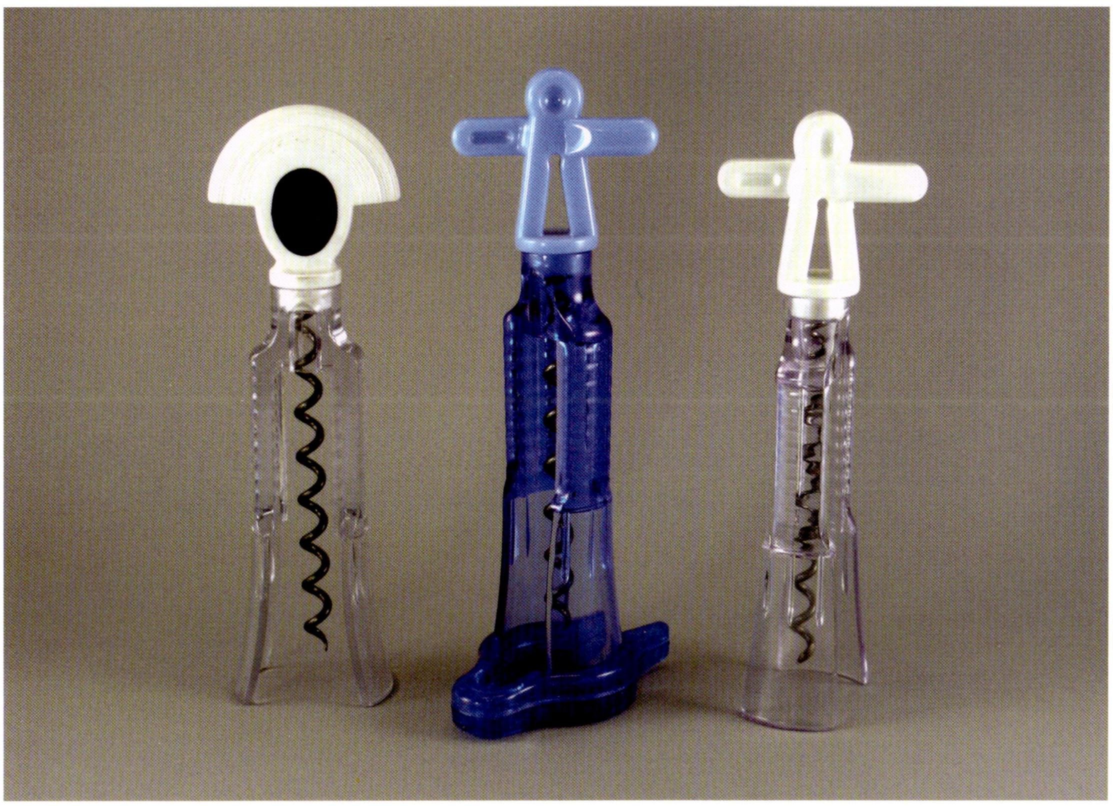

Unknown plastic corkscrews. All have the same frame, different tops, and the center one has a stand/foil cutter.

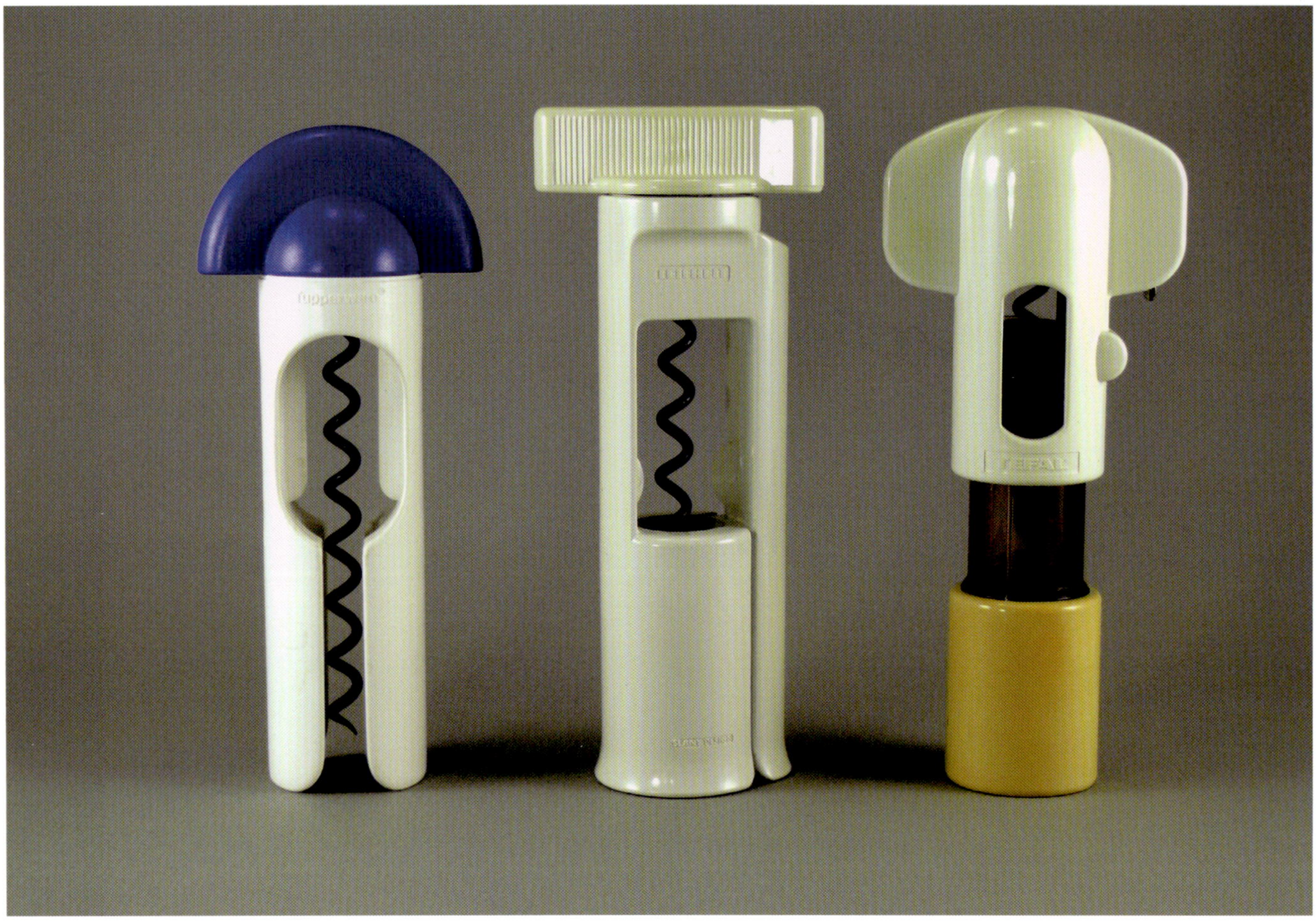

Left to Right: Tupperware; Leifheit, Germany; and Starfrit + Tefal, France.

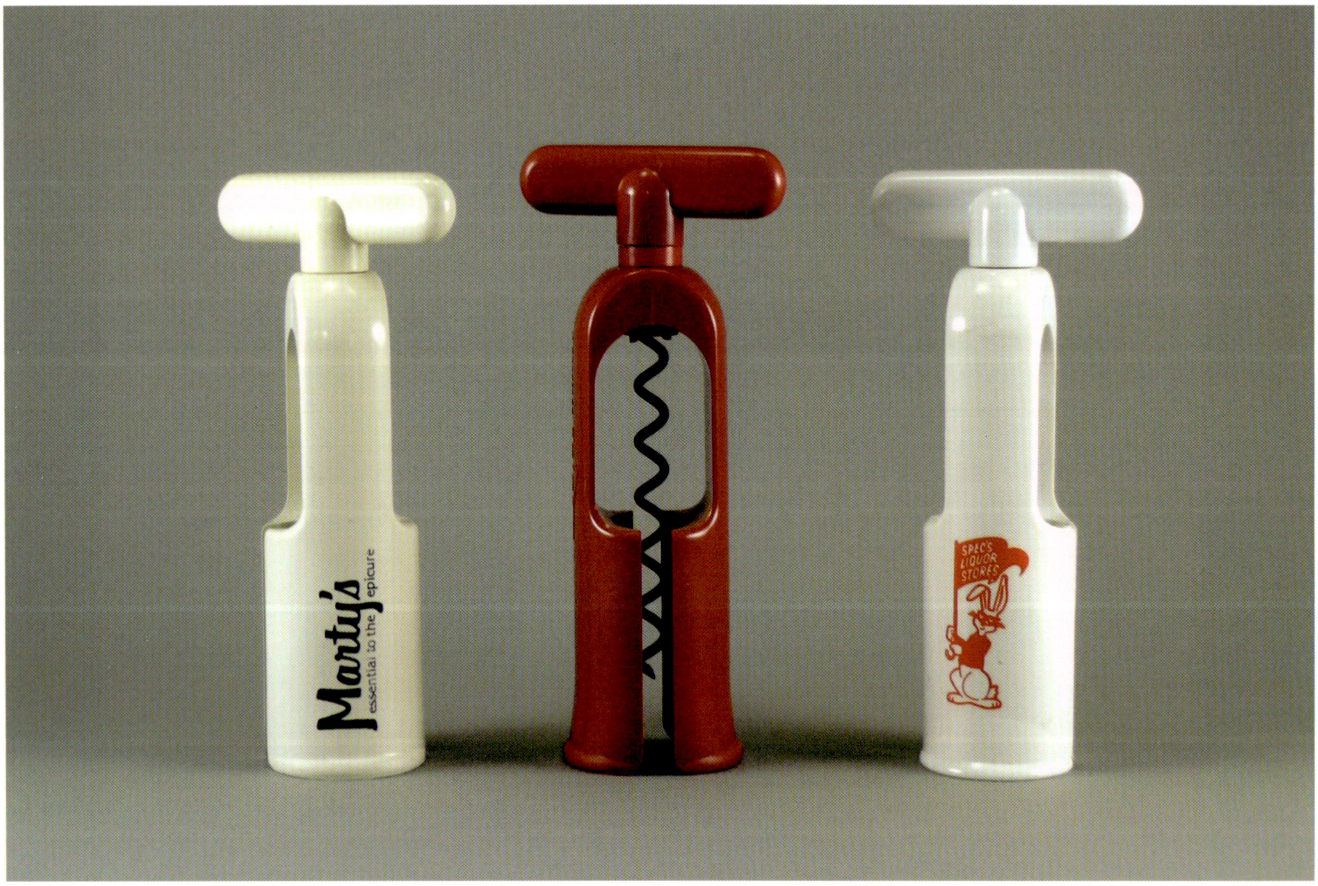

Center: Snips, il Cavatappi, Italy. The white ones are also Snips with business logos printed on them.

Bodum, designed and manufactured in Portugal. All marked on inside of one leg.

The OXO, a later version of the Xyliss, and a Lafeith.

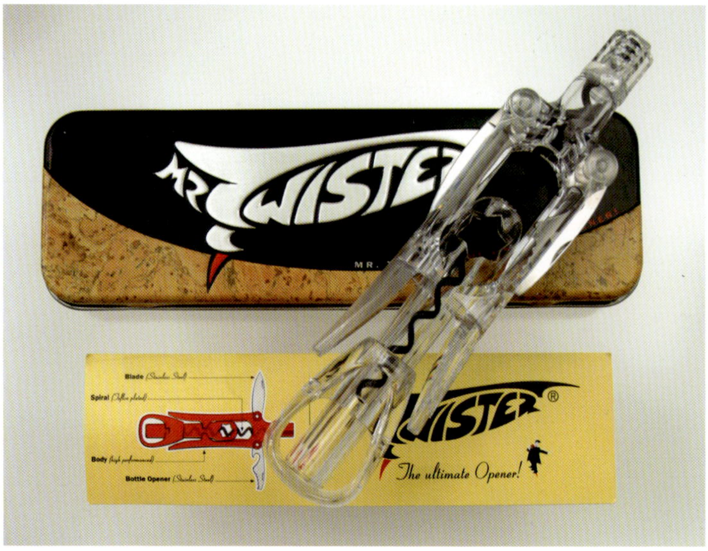

Mr. Twister spinhandle pocket corkscrew with knife openers.

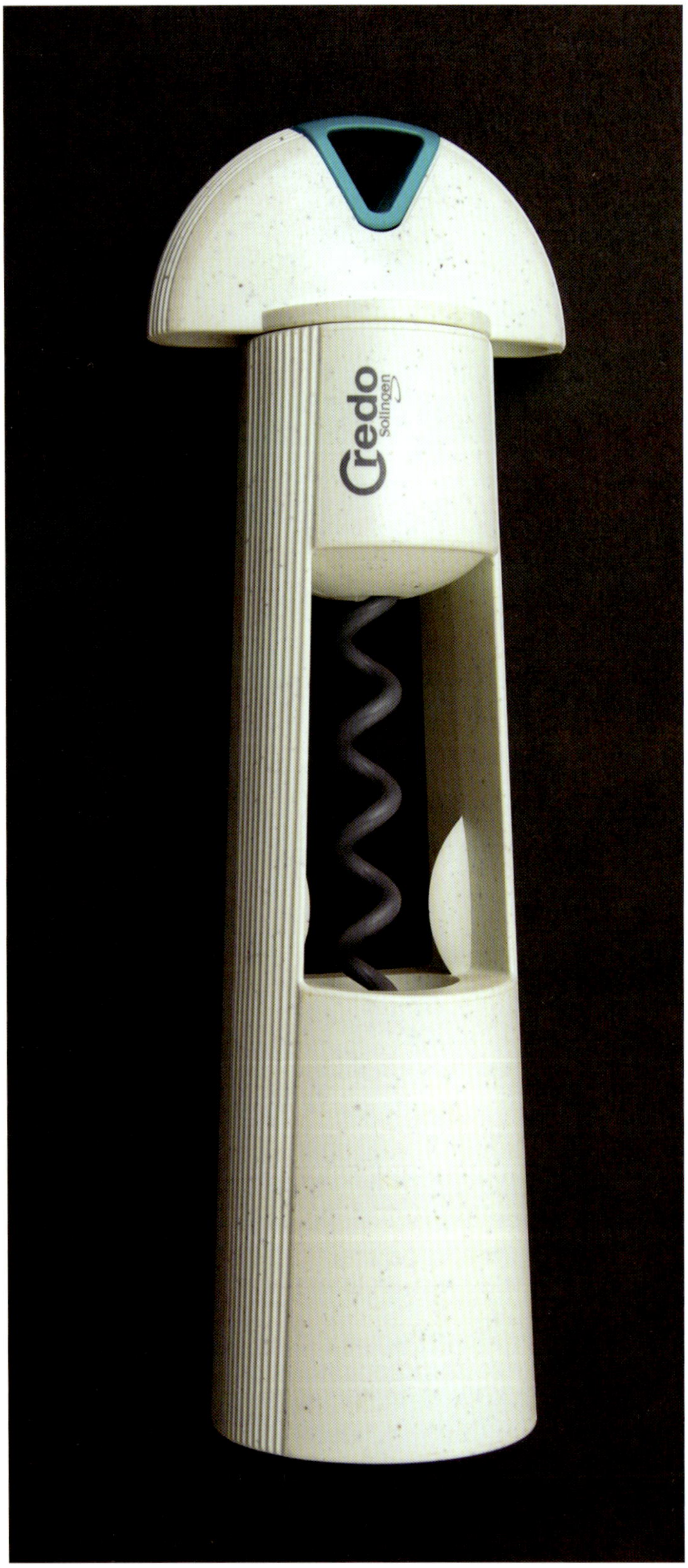

The Credo, by Solingen, Germany.

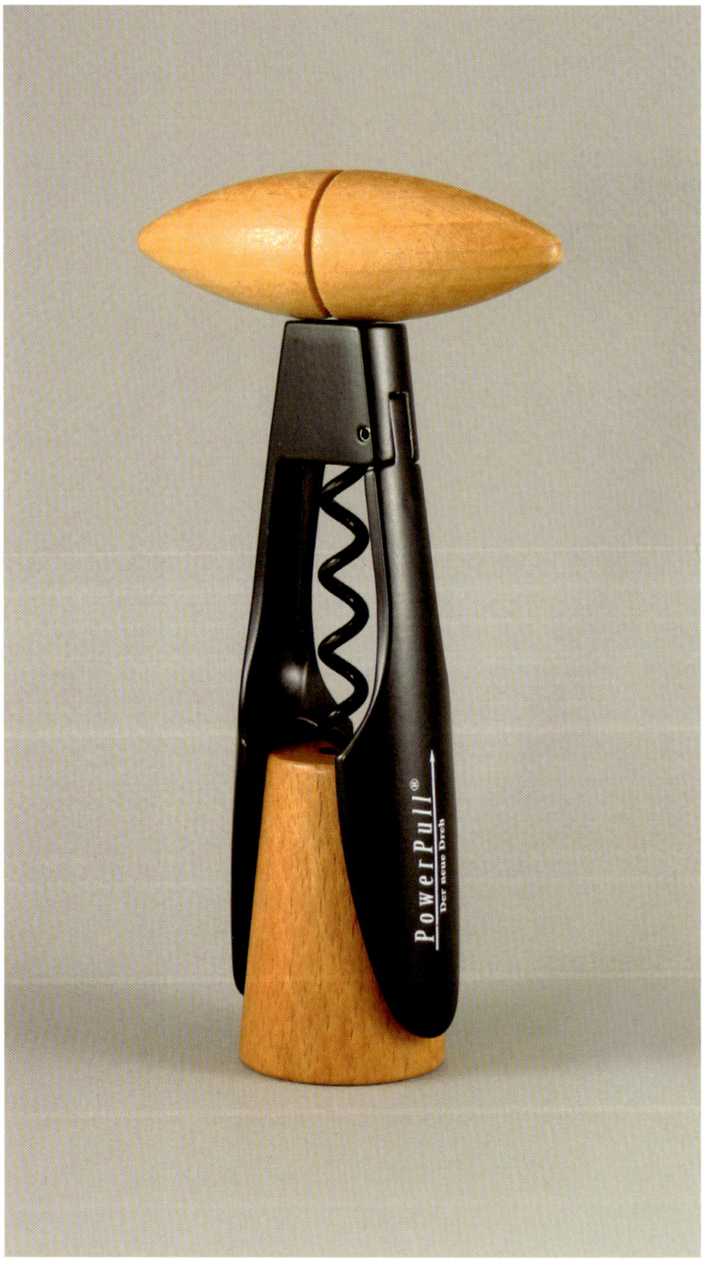

PowerPull, Germany.

The Largest Screwpull Knockoff in the World

If you have ever visited the White Fences Winery in Irvington, Virginia, you have seen two of the largest "Screwpulls knockoffs" in the world. They adorn the entrance to what was once the White Fences Winery. Unfortunately, the winery went out of business in 2010 and was sold to the Hope and Glory Inn. The vineyard and the corkscrews are still there and are maintained by the new owners, Dudley and Peggy Patterson.

The corkscrews are each 25 feet tall and patterned after the Screwpull Table Model. The corkscrews were commissioned by Bill Wadsworth, owner of White Fences at the time. They were fabricated in 2002 by Tiffany Yachts, just a short distance away in Burgess, Virginia, on the Chesapeake Bay. According to Randy at Tiffany Yachts, they were built with a steel internal structure and the surfaces were fiberglass, built up like you would build a yacht. They are really a sight to see and today show the way to the Hope and Glory Inn.

Largest Screwpull knockoff in the world, White Fences Winery, Virginia. *Courtesy Sherrill Yost*

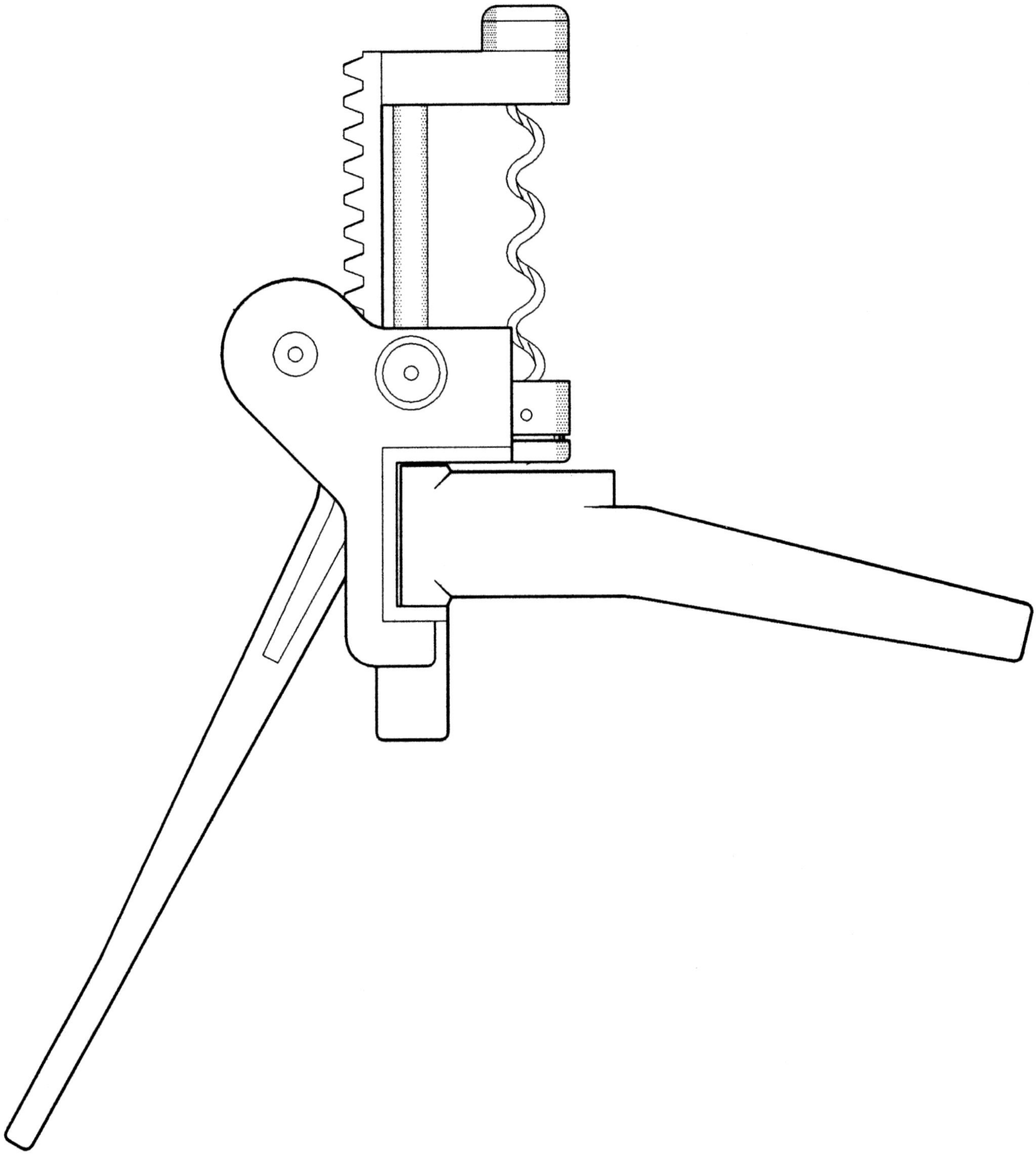

Chapter 5: The Leverpull Prototypes

Herb Allen ran many more calculations and made many more prototype screws in his effort to come up with a pressure-applied screw for his Leverpull idea. The successful development of the Screwpull screw led Allen into engineering a pressure-applied screw in lieu of a turning force that could be used in a lever-type machine. Allen's calculations developed an elongated pitch version of the Teflon-coated screw, with a thrust bearing at the top. The elongated pitch of the screw and the nut/bearing assembly device required for direct pressure insertion were satisfactorily solved fairly early in his invention process for the Leverpull. The Leverpull Model became a reality.

Herb Allen started building prototypes of his idea for the Leverpull. These prototypes utilized a system of levers and pins with offset and moving fulcrums to keep the "pull" of the cork in a straight line. The early prototypes all worked, but the action was not as smooth as he thought it should be, so he kept trying, with his trial-and-error methods, changing lever lengths, fulcrum points, etc. But still he was not satisfied with the feel and the problem prone areas in the design, possibly reducing the longevity of the corkscrew.

The prototypes worked well enough for Allen to go ahead and apply for a patent on the Leverpull. Drawings were produced using two versions of the prototypes for the application. On March 3, 1981, U.S. Patent Number 4,253,351 was issued to Herb Allen and the Hallen Company.

Herb was still not happy with the design. Production complexity worried him. Too many things could go wrong after many cork extractions: this was of great concern to him. Cork removal from the screw was also giving him problems, and he felt that this really needed improvement. After many discussions with him, it seems that prototypes #6 through #8 were all about trying to solve the problem of cork removal from the screw. The empty frame shown on page 64 are all that is left of prototypes #6 through #8. He made new and different parts to solve this problem and change them out with this frame. Apparently, prototypes #6 through #8 were abandoned as not being viable working ideas, and this frame is all that is left of these screw-drive assemblies.

This photo shows a lever-type prototype, assumed #6 through #8, without the experimental parts trying to improve the system of cork removal from the worm.

This version of the Leverpull, with the levers and fulcrums, as shown in the patent drawings (pages 108–111), consumed much of Herb Allen's time trying for perfection. Prototype #1 is the one he showed in his cellar workshop in 1979.

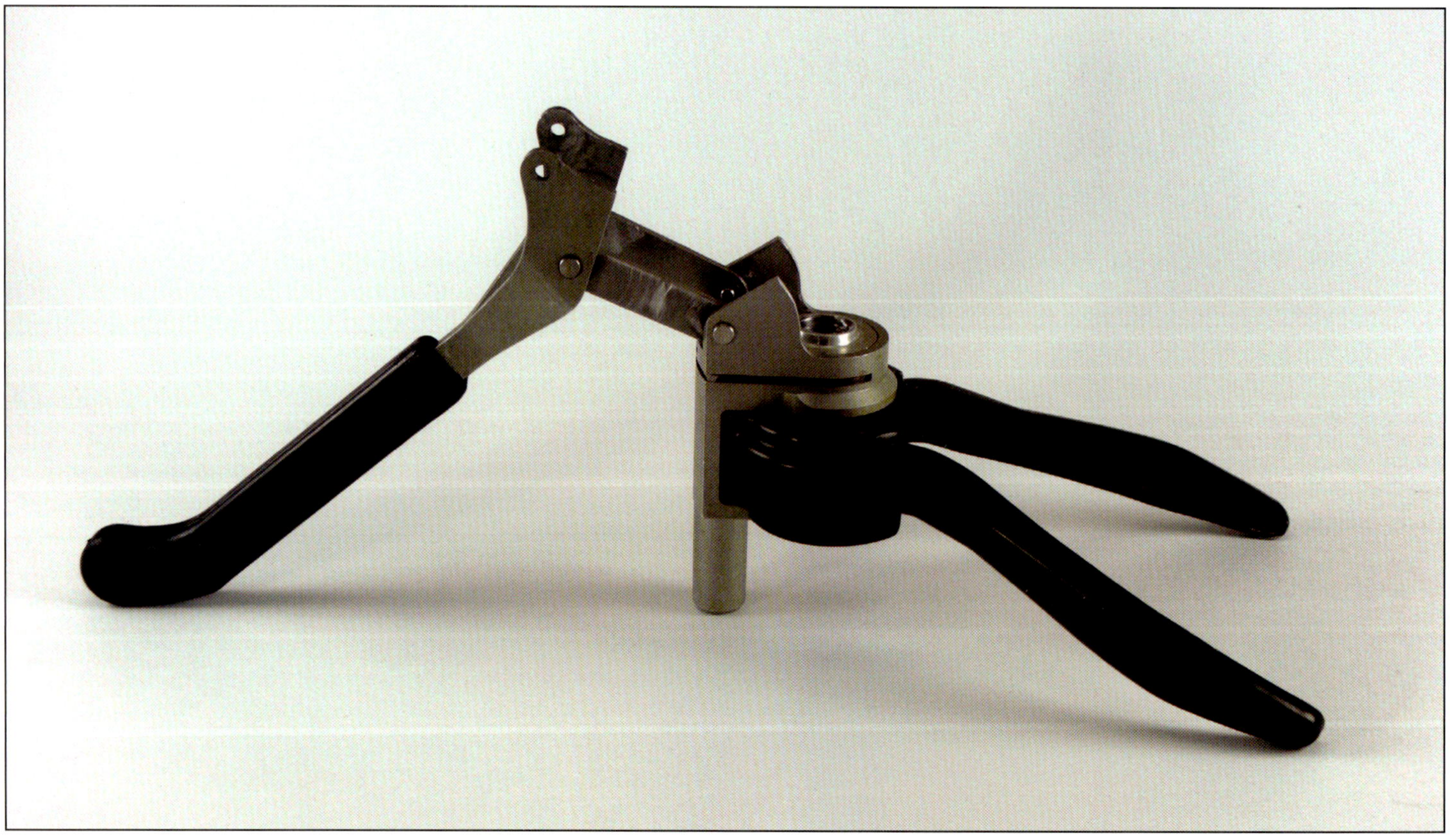

Leverpull Prototype #6 through #8 body.

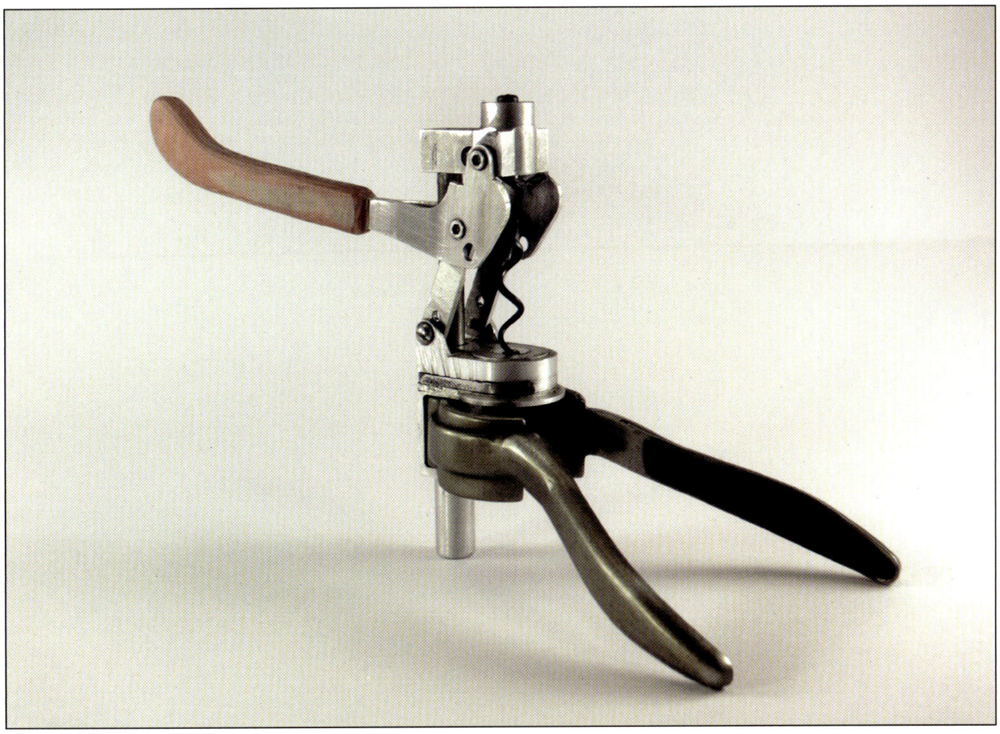

Lever type Screwpull Prototype #1 in the "open" position.

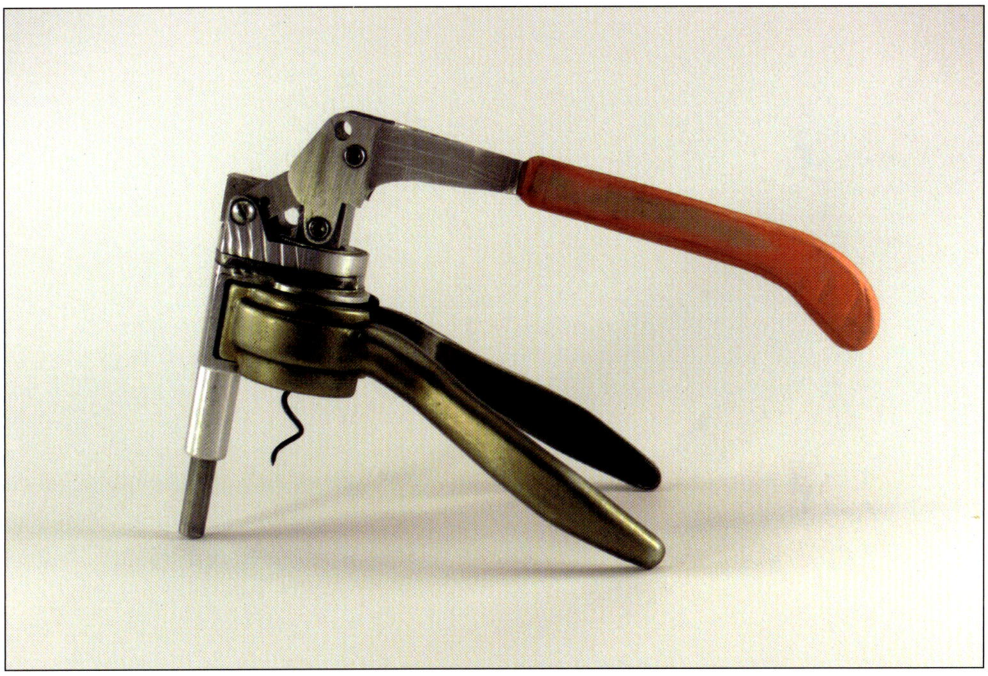

Lever type Screwpull Prototype #1.

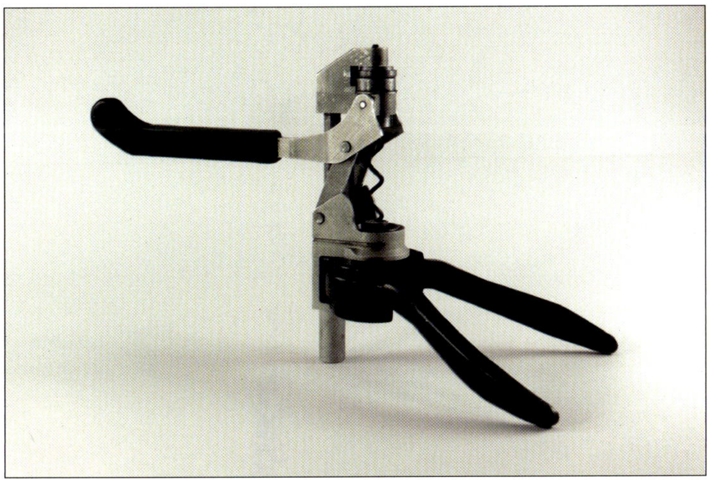

Lever type Screwpull Prototype #9 in the "open" position.

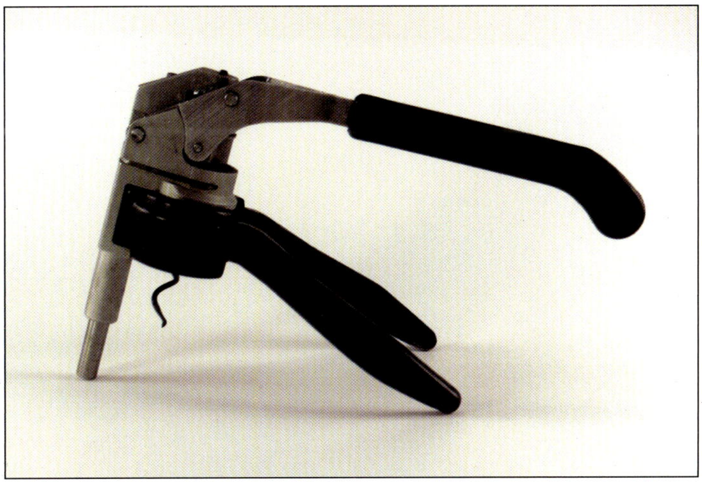

Lever type Screwpull Prototype #9.

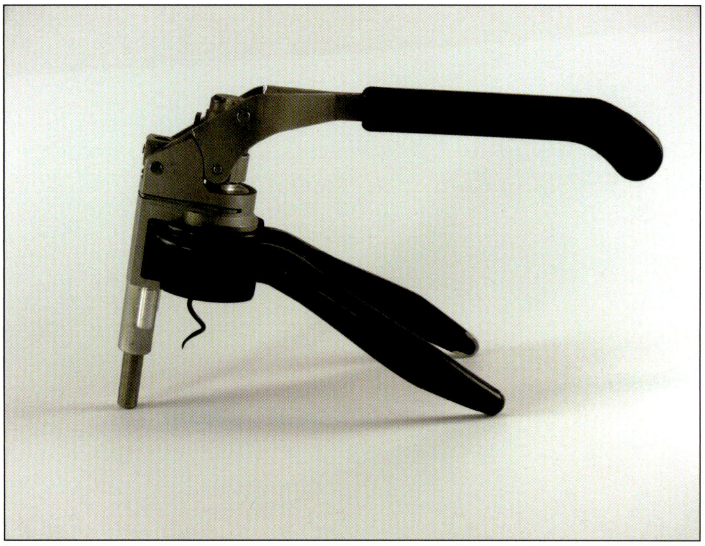

Lever type Screwpull Prototype #10.

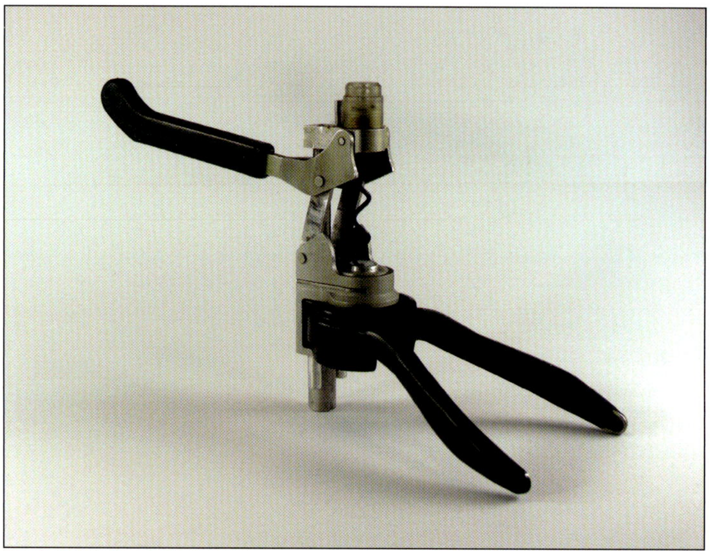

Lever type Screwpull Prototype #10 in the "open" position.

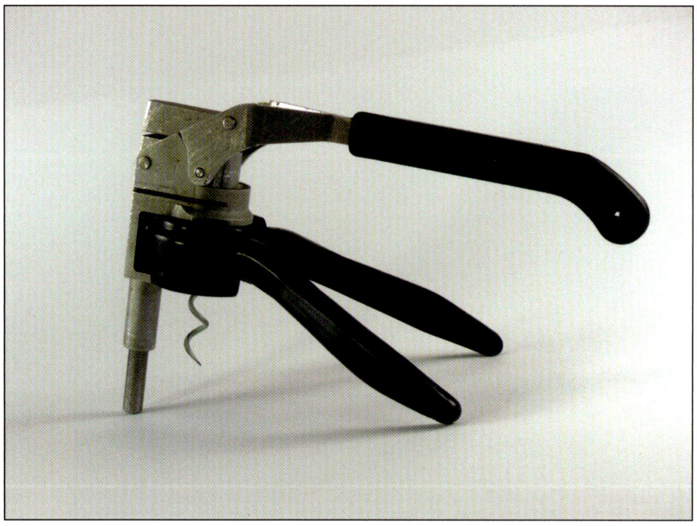

Lever type Screwpull Prototype #11.

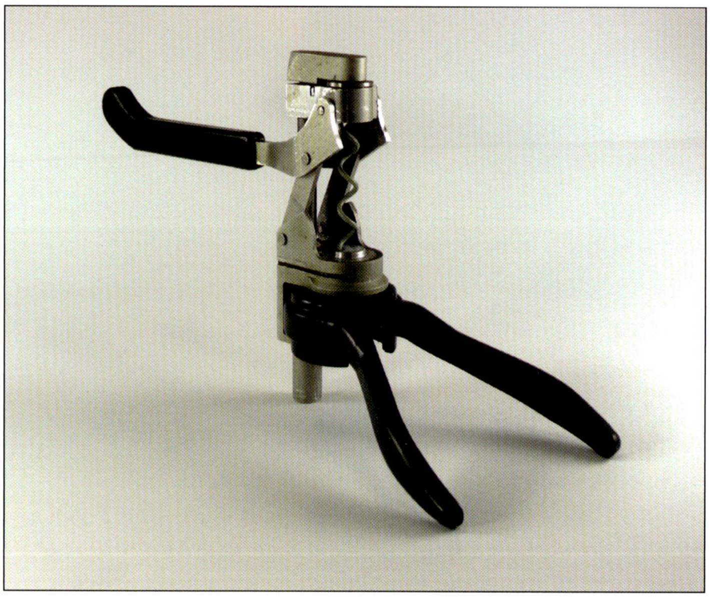

Lever type Screwpull Prototype #11 in the "open" position.

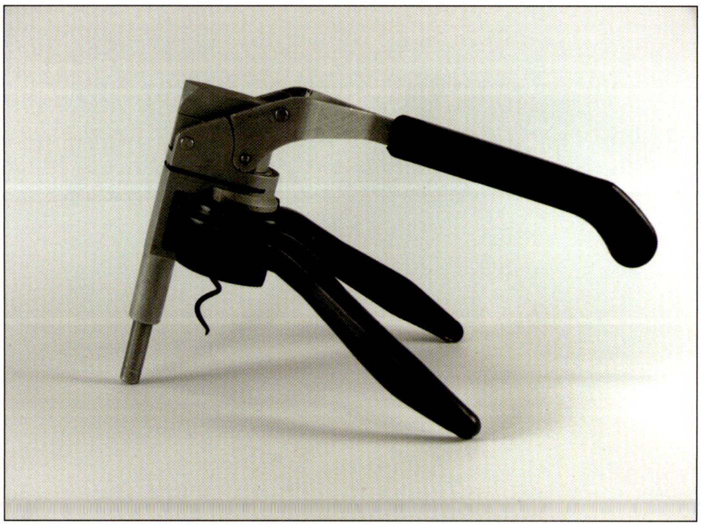

Lever type Screwpull Prototype #12.

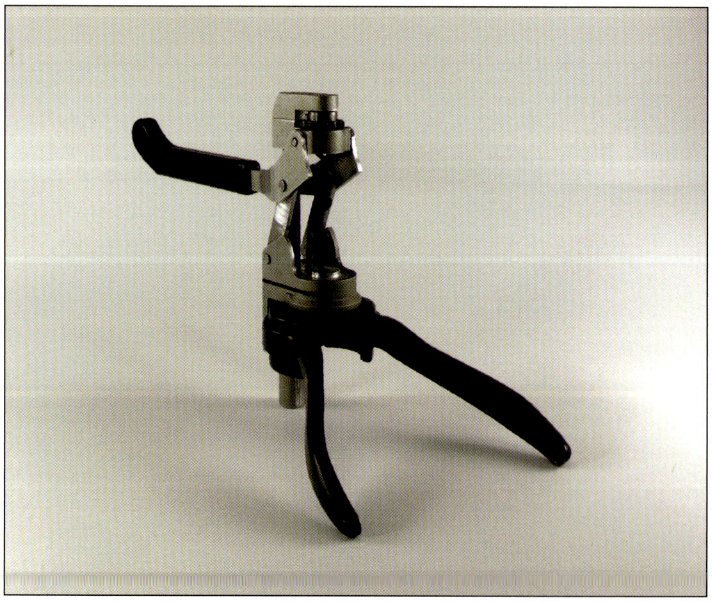

Lever type Screwpull Prototype #12 in the "open" position.

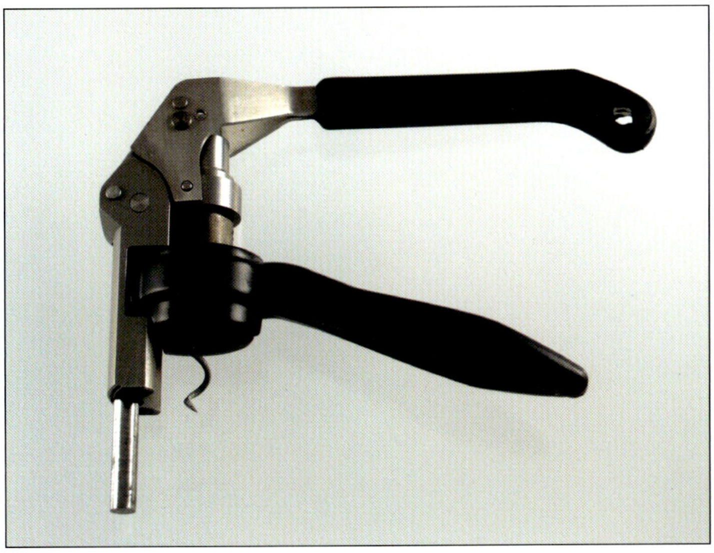

Lever type Screwpull Prototype #13.

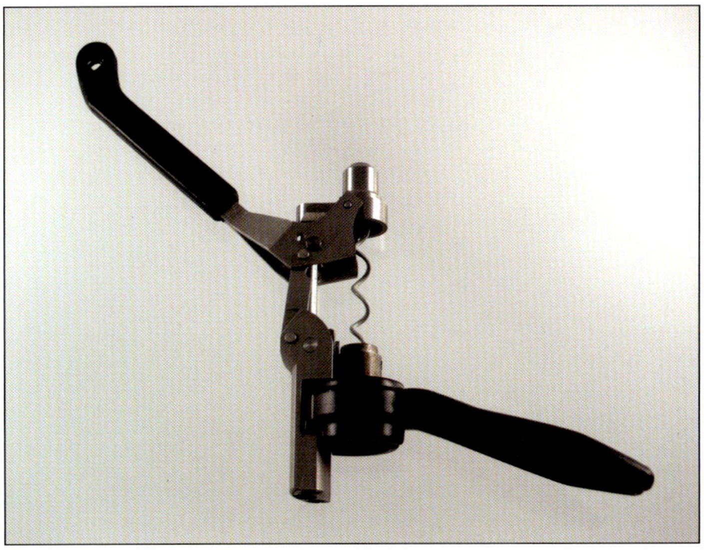

Lever type Screwpull Prototype #13 in the "open" position.

The six prototypes shown here, #1 through #13, show the progression of improvements and Herb Allen's trial-and-error methods of accomplishing a solution to an idea. Number 13, the final version of this design, is the one that Allen used for his patent, patent drawings, and patent descriptions. Number 11 was also described in the patent and sheets 5 through 9 of the patent drawings.

Note that the bottle-grip handles did not change in all of the prototypes, except that #1 has a shiny exposed metal finish where all of the others have a black coated finish. The body is basically the same on #9, #10, #11, and #12. Number 1 is fundamentally the same as all of the others except #13, which is totally different. The construction of the pull lever on all of the prototypes vary only slightly to fit the lever fulcrum points and the body shape. The cast rubber over the metal handle of #1 is orange in color, where all of the others have a black rubber material over the metal.

The screw in #11 and #13 is white, indicating that Allen was experimenting with different Teflon coatings at this early stage.

The mechanisms at the screw head and screw guide systems are all different.Allen tried to find a reliable solution to raising the screw guide and removing the cork from the screw. They appear to be similar on the surface, until you operate them and closely observe how the actions work, revealing the functions of each moving part. Then you see how different they really are.

Lever type Screwpull Prototype #11 was used for the prototype in the March 3, 1981 Patent #4,253,351, Sheets 6 through 9 of the drawings (see pages 108–111).

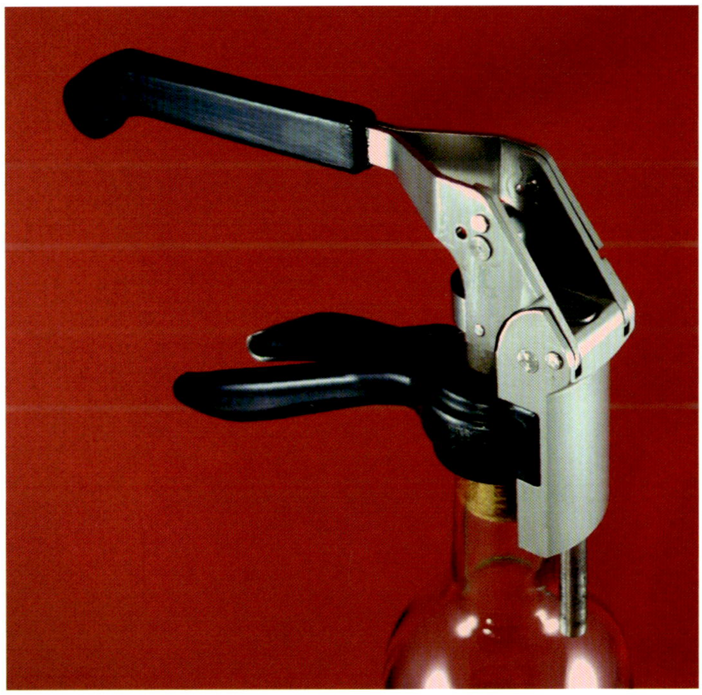

Lever type Screwpull Prototype #13 was used for the prototype in the March 3, 1981 Patent #4,253,351, Sheets 1 through 5 of the drawings (see pages 108–111).

Some of the ideas from this lever system type are also used in the final production models of the rack-and-pinion gear style corkscrews described below.

Rack-and-pinion Leverpull Prototypes

Herb Allen was thinking, "There has to be a better way of doing it." His drive for perfection led him into a totally different direction for the Leverpull. He shifted from the lever style of mechanical advantage to a rack-and-pinion style, still with a lever as the driving force. He could see long-range problems with the six points of pins connecting the lever arm to the fulcrum arms to the body. Possible premature wear and still not having a smooth withdrawal of the cork prompted Allen to find "the better way." Gears were proven to be a very dependable and trouble-free way of providing a mechanical advantage, so he took the road toward a rack-and-pinion gear system.

Almost immediately, from prototype #1 on, he had success. Each prototype after #1 became an improvement over the last one he made. The latches, that hold the screw guide down while pulling up on the lever arm, became a viable solution to removing the cork from the screw. This was a problem with the lever-style system from the very beginning.

In order to remove the cork from the screw, or the screw from the cork after it has been pulled from the bottle, you have to return the lever arm back to the closed position and pull the lever arm up, which pulls the screw out of the cork. The action of releasing the pressure on the bottle-grip handles, after removing Leverpull from the top of the bottle and returning the lever arm back to the closed position, locks the latches over the screw guide holder, keeping it from returning to the top. In turn, this allows the screw to unwind out of the cork while pulling up on the lever arm.

The bodies and bottle-grip handles, along with some of the lever-arm pinion gears, on all of the first nine or so prototypes, were sand-cast out of aluminum. These parts were machined and then assembled with the other individually machined parts.

Prototype #9 was so successful that Allen proudly demonstrated it at the Right's Reception of the 1984 Annual General Meeting of the International Correspondence of Corkscrew Addicts at Baileys Hotel in Kensington, London. Shortly after, the composite material prototypes were cast, assembled, and tested. They also were successful so that they went into production almost immediately and the Leverpull became an overnight success story.

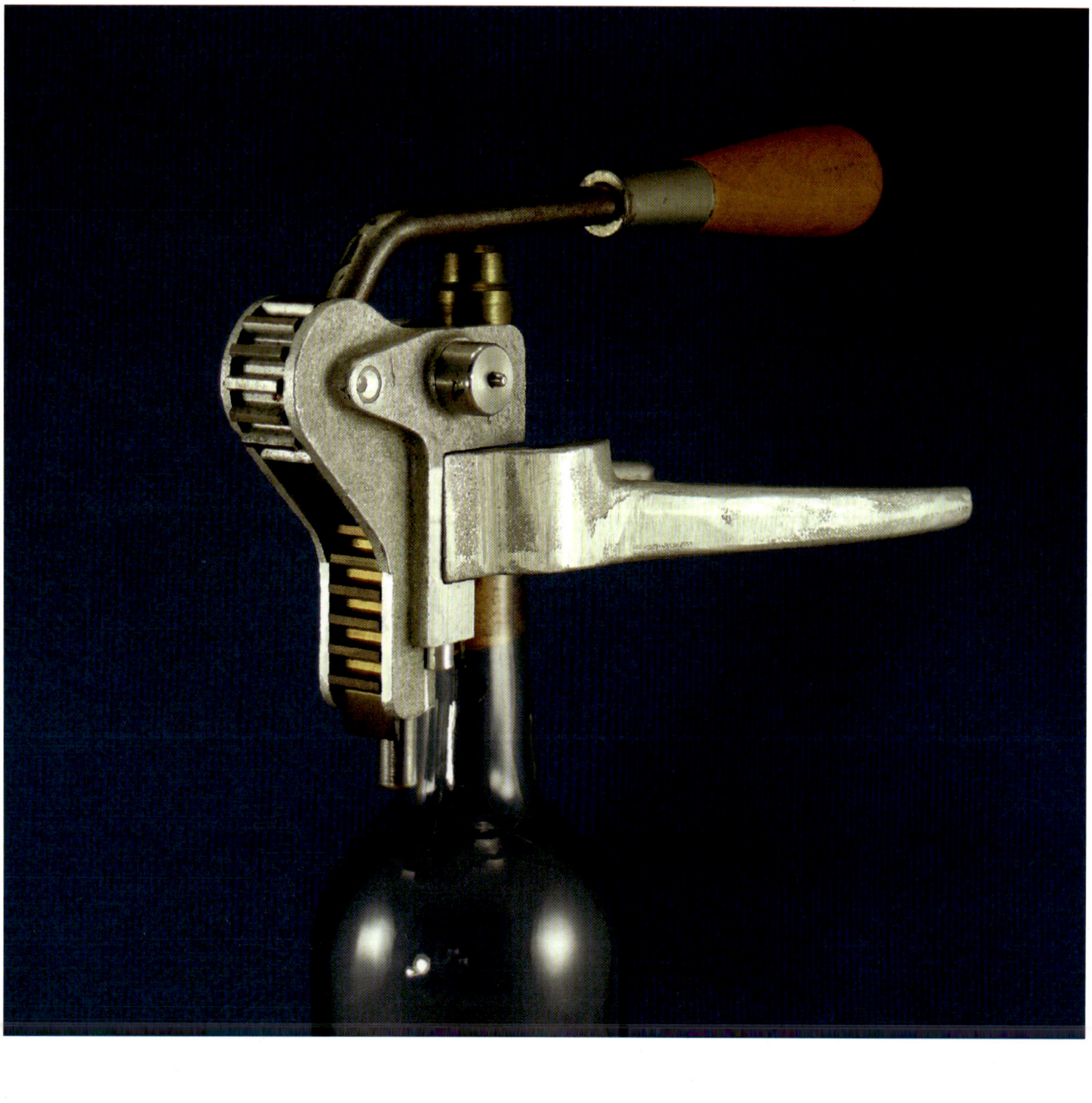

The following pages feature six rack-and-pinion gear prototypes and their explanations.

Prototype #1

The two photographs on this page are of the first prototype of the rack-and-pinion, gear-style Leverpull. The main body and bottle-grip handles were sand-cast out of aluminum and then machined to receive the hinge pins, pinion gear, etc. The rack and screw-drive housing was machined from solid brass bar stock and the pinion gear was machined from steel. What to use for a handle or pull lever? A grill fork was commandeered. The fork was surgically removed, the 3/8" round rod was threaded, then bent to shape and screwed into a threaded hole in the pinion gear. It works quite well.

The rack-and-pinion gear are a full 5/8" wide, with full-width teeth. The unit is 9 1/2" long and weighs over three pounds. This is quite a hefty machine.

The bottle-grip handle's pivot pins are inserted from the

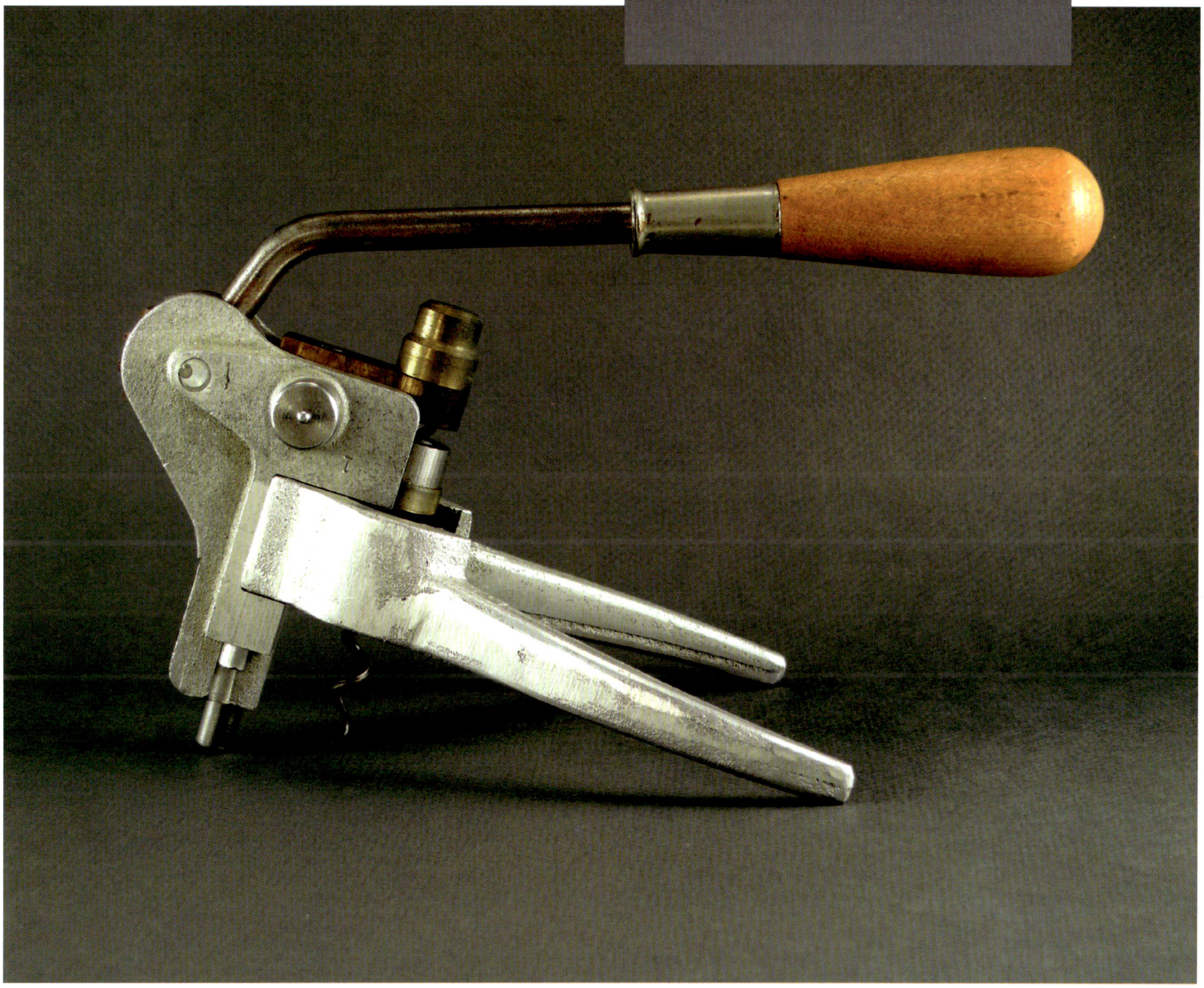

Prototype #2

The second prototype of the rack-and-pinion gear-style Leverpull body and bottle-grip handles was also sand-cast from aluminum and machined the same as #1. The pull lever handle was integrated with the pinion gear and also sand-cast from aluminum. The gear was machined to fit the rack. The rack was machined from solid brass bar stock.

The rack-and-pinion gear are also 5/8" wide, with full-width teeth. The unit is 9" long. The bottle-grip handle pivot pins are inserted from the bottom, and the bottle-grip handles are 6" long.

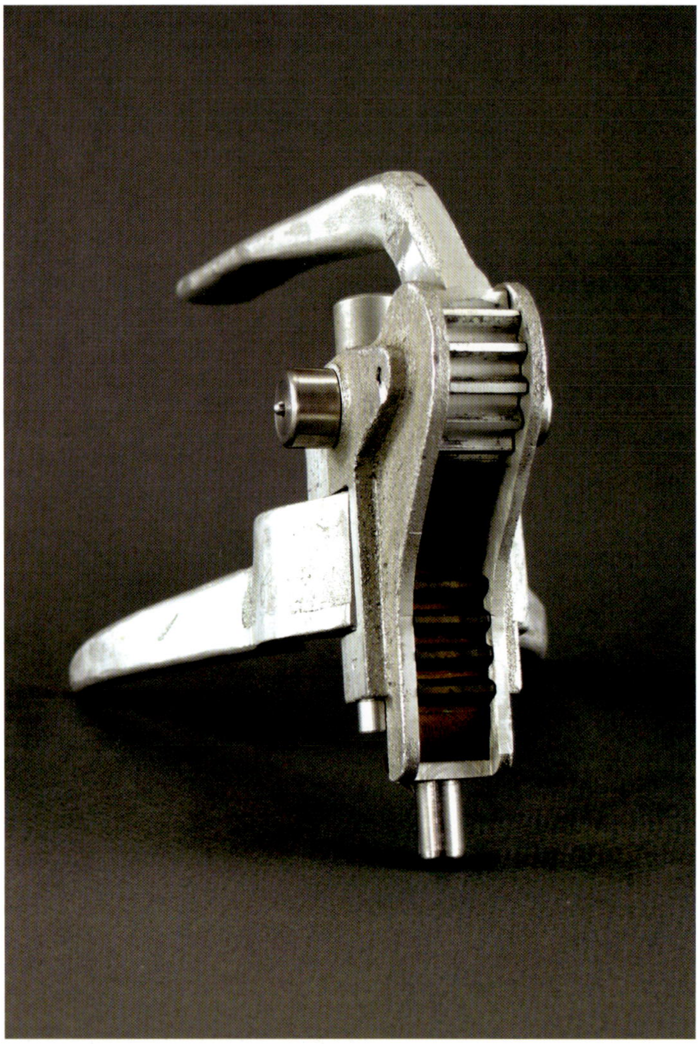

Prototype #3

The third prototype of the rack-and-pinion gear-style Leverpull body, bottle-grip handles, and the pull lever were sand-cast from aluminum and machined like #2.

The rack, with the screw drive housing, on #3 is machined from aluminum bar stock instead of the brass of #1 and #2.

The rack-and-pinion gears are 5/8" wide, with full-width teeth. The unit is 9" long.

The bottle-grip handle pivot pins are inserted from the bottom and the bottle-grip handles are 6" long.

Prototype #5

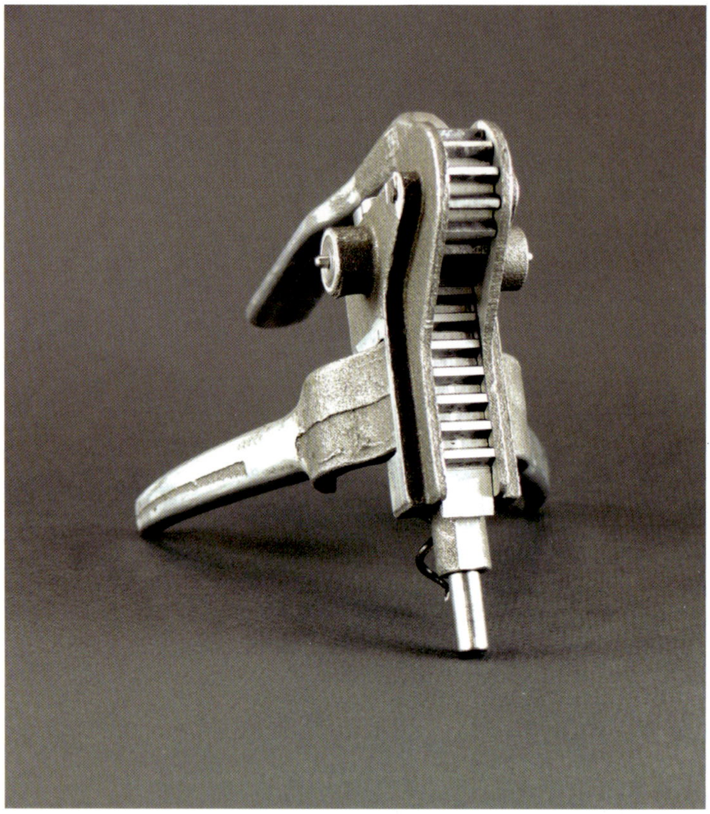

Prototype #5 of the rack-and-pinion gear-style Leverpull body, bottle-grip handles, and the pull-lever with integrated pinion gear were sand-cast from aluminum and machined, much like the previous units. The rack and screw-drive housing is machined from solid aluminum bar stock.

The difference between #5 and the previous three prototypes is size. The rack-and-pinion gear are 1/2" wide, instead of 5/8". The unit is 8 1/8" long, instead of 9", and the bottle-grip handles are 5 3/8", instead of 6". The bottle-grip handle pivot pins are inserted from the top.

This unit is nearly the size of the production models, but still has the triangular projection out to the pull-lever pin axle.

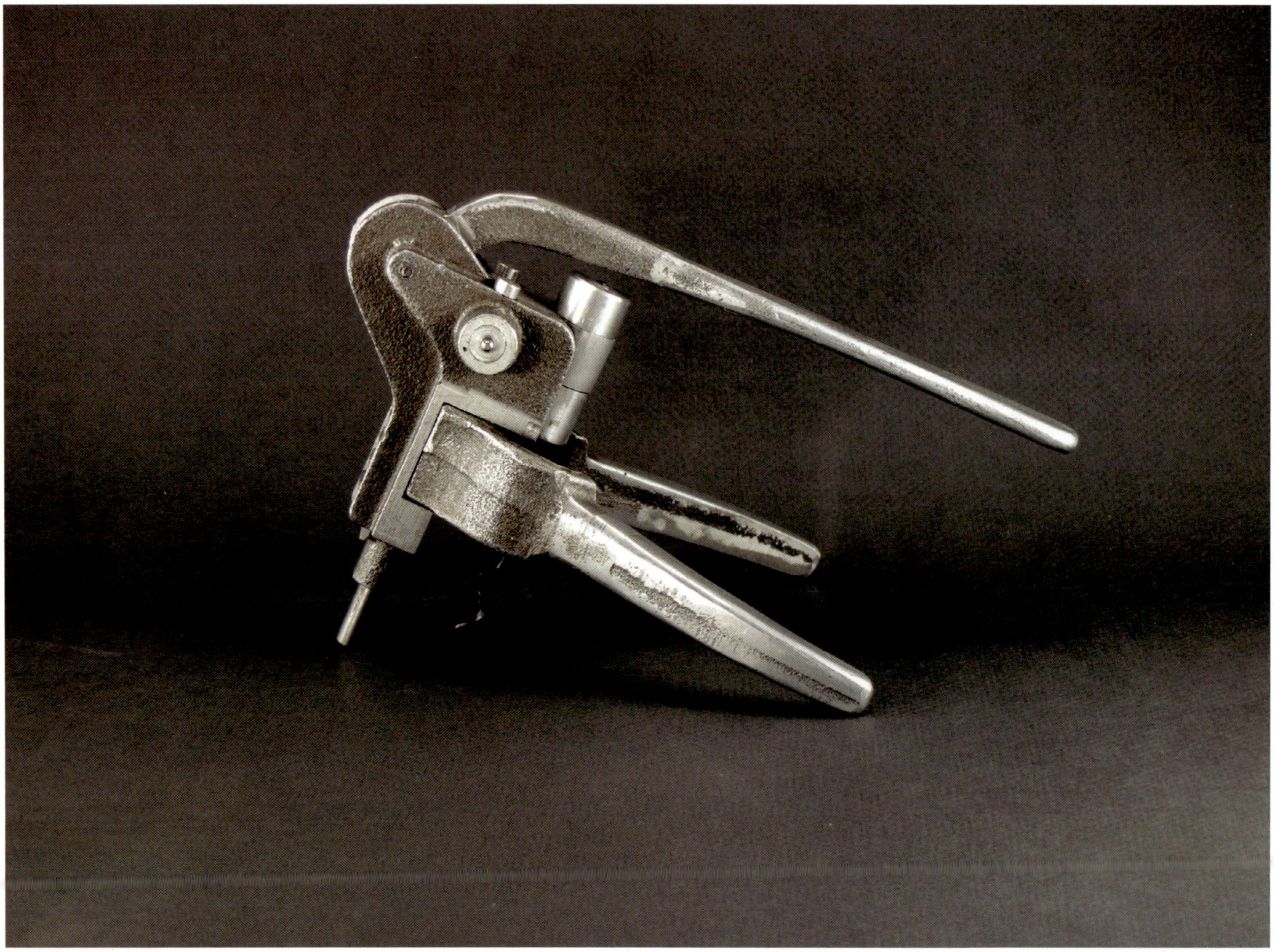

Prototype #9

Prototype #9 of the rack-and-pinion gear-type prototypes was the last cast aluminum prototype made. It is the nearest aluminum prototype in size and shape to the final production models. Note that there is no triangular projection out to the pull-lever axle pin. The pull-lever and bottle-grip handles are similar to the final production models. The size is the same as the final production models, which is smaller than all the other aluminum prototypes.

The bottle-grip handle pivot pins are inserted from the bottom.

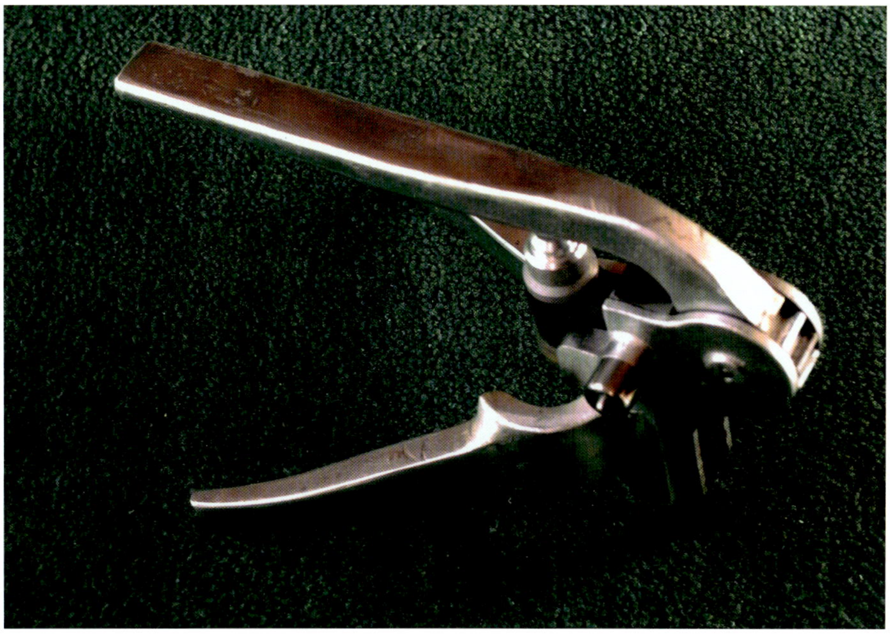

Courtesy Dr. David Bradshaw.

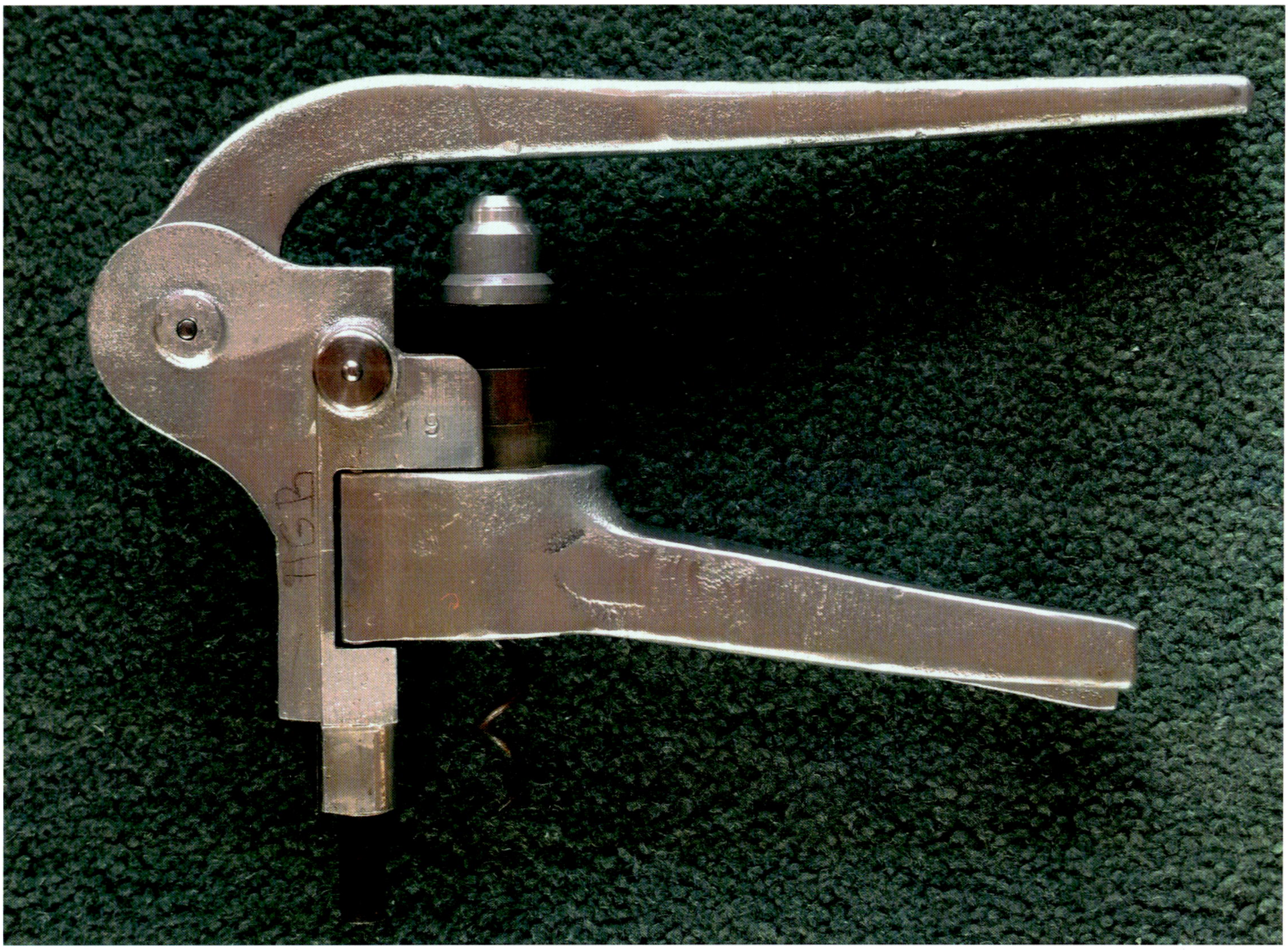

Courtesy Dr. David Bradshaw.

Left to Right: Start-Right Dr. Bernard M. Watney (ICCA member and Chief Medical Officer of the Guiness Brewery, London); Herbert Allen (Honorary ICCA member and CEO of the Hallen Company, Houston); and Just-Right Brother Timothy Diener (ICCA member and cellar master of Christian Brothers' Winery, Napa Valley) demonstrating Leverpull Prototype #9 at the 1984 International Correspondence of Corkscrew Addicts Annual General Meeting in London on some select Christian Brothers' wine. *Courtesy of Herb Danziger.*

This prototype is quite unusual in that the body, grip handles, and pull-lever are constructed of a graphite/fiberglass composite material much like the final production models. You may note that this prototype has a flat-bladed screw. Allen thought that the flat blade might be stronger for a direct pressure insertion than a round wire. His comment was, "I had to try it. It had the strength but it did not run through the guide nut as smoothly as the round one."

The rack-and-pinion are 5/8" wide, the unit is 7 1/2" long, and the bottle-grip handles are 5 1/4" long, with the hinge pins inserted from the top.

Also note that the "Screwpull" logo is not cast into the flat face area around the pull lever axle pin, as on the production models, and the screw drive and rack metals are not a finished metal.

This prototype model is the first one where Allen used the split-tooth rack-and-pinion gear. There are no identifying marks.

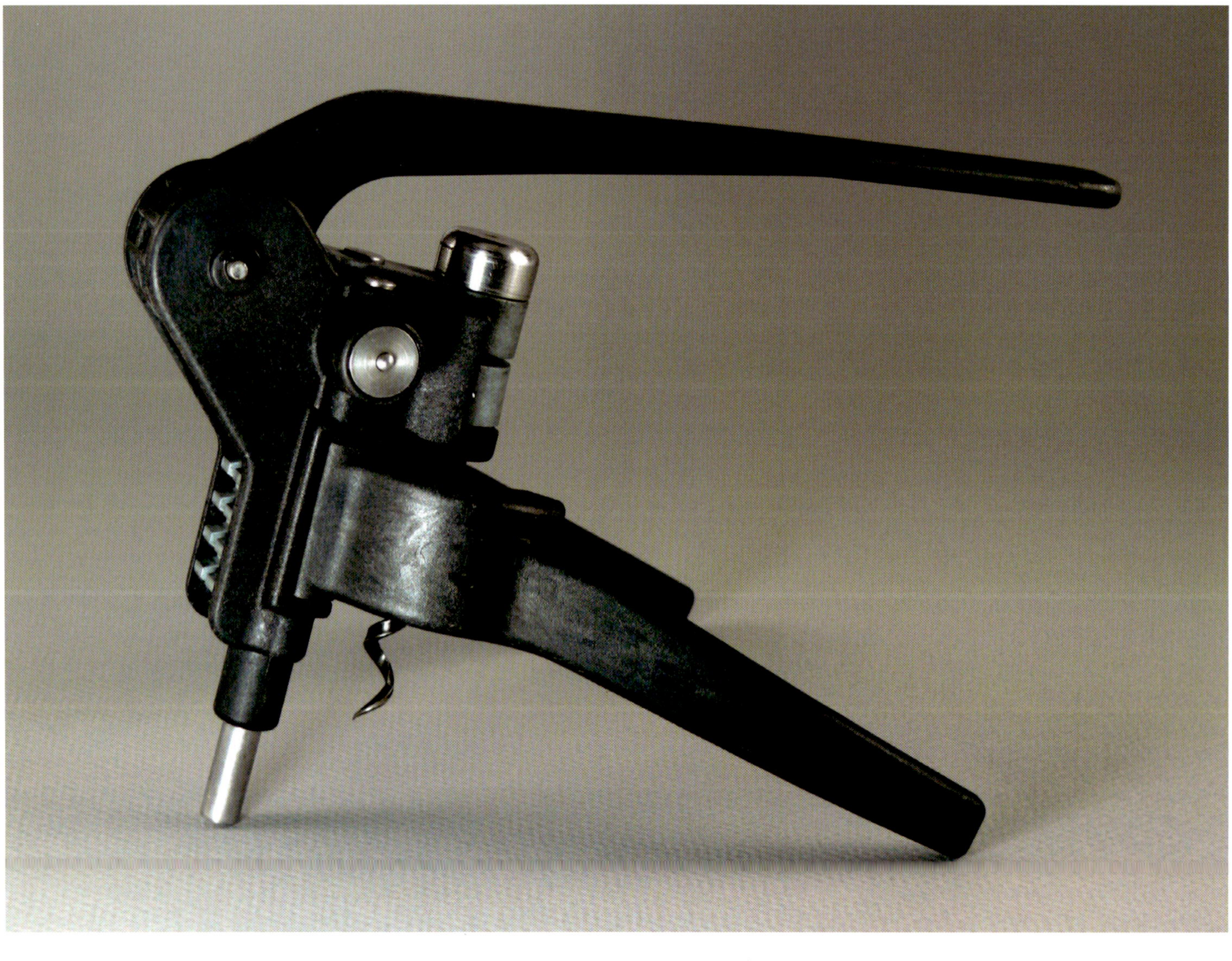

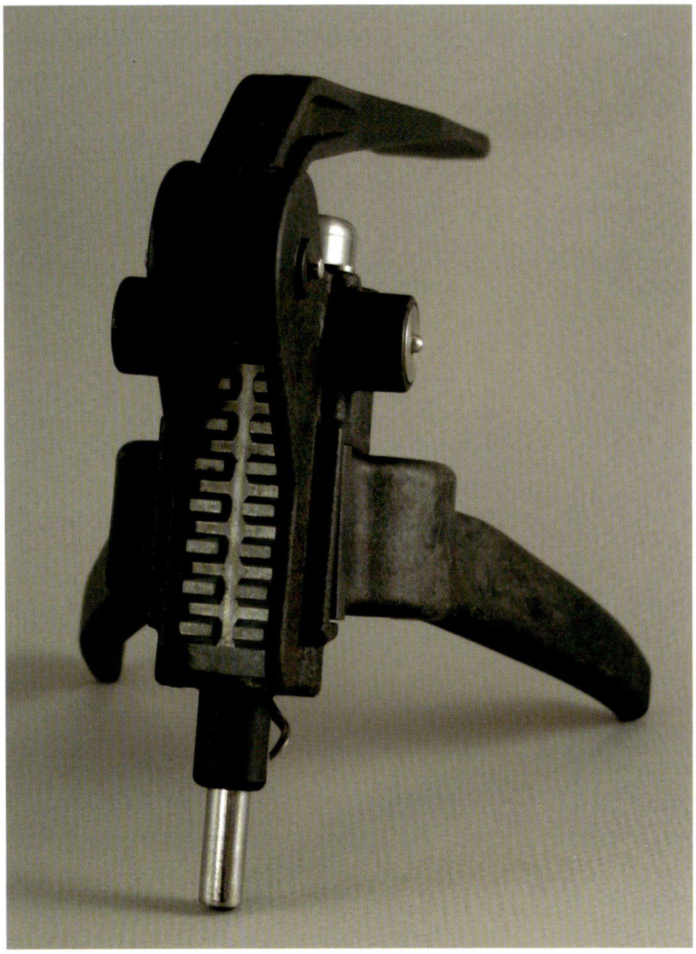

The last prototype made before production was launched.

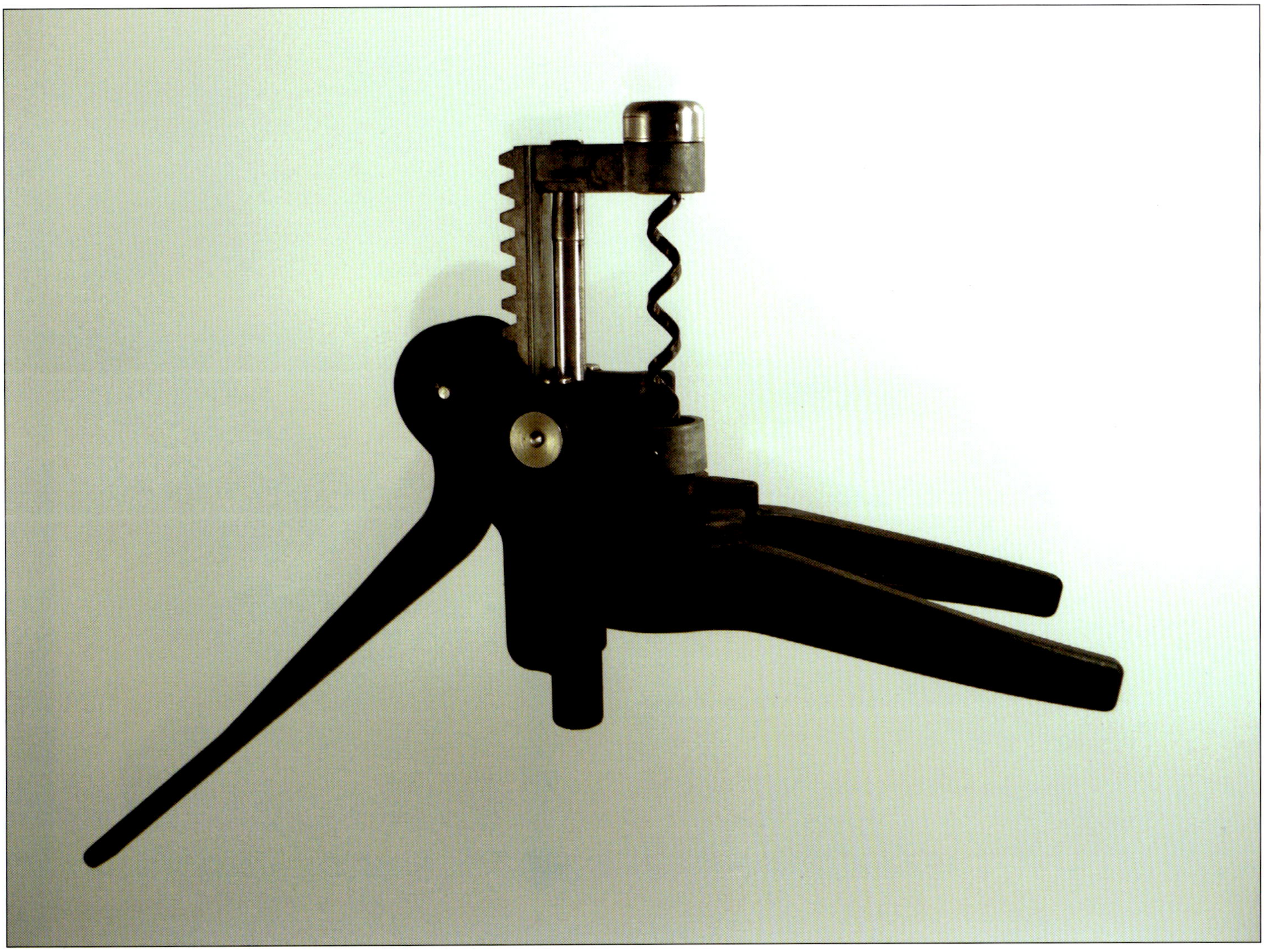

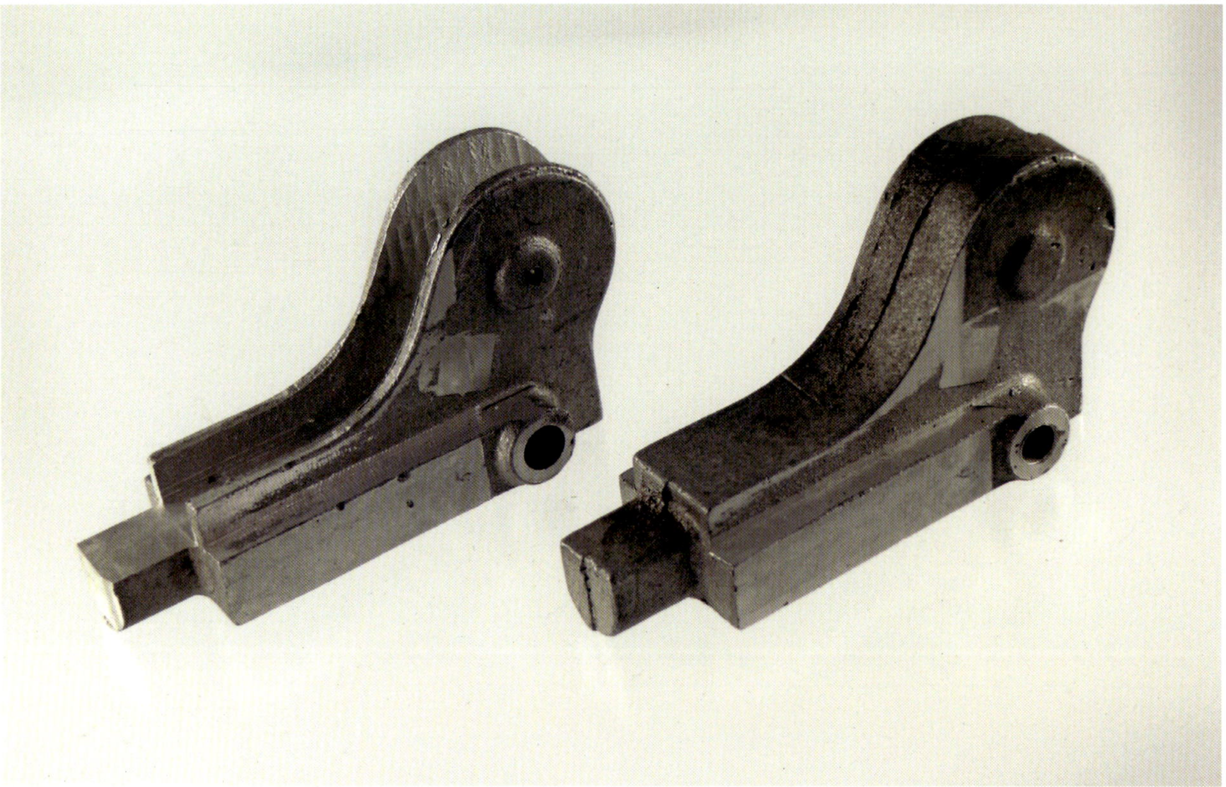

Leverpull body parts

The two photos above show Leverpull body parts in various stages of machining. The top left piece shows an almost raw aluminum casting with only the latch pin area hole drilled. The top right piece has the latch pin area hole drilled as well as the area for the rack-and-pinion gear machined out of the raw casting. These castings are for the later prototypes, such as #9. Note that there is no triangular-shaped raised shoulder in the pull lever pivot pin area, as in the earlier versions of prototypes such as #1 through #5.

The lower photo shows two graphite/fiberglass composite material body castings, either rejects for more prototypes or to be used on actual early production models. The logo with the word "LEVERPULL" was used only for a very short time.

Close-ups of the two body parts showing the two different logos that were used on the Leverpulls. The one with the words "LEVER PULL" was used only on very early models, with hand-stamped serial numbers in the handle, prior to being machine-stamped behind the rack.

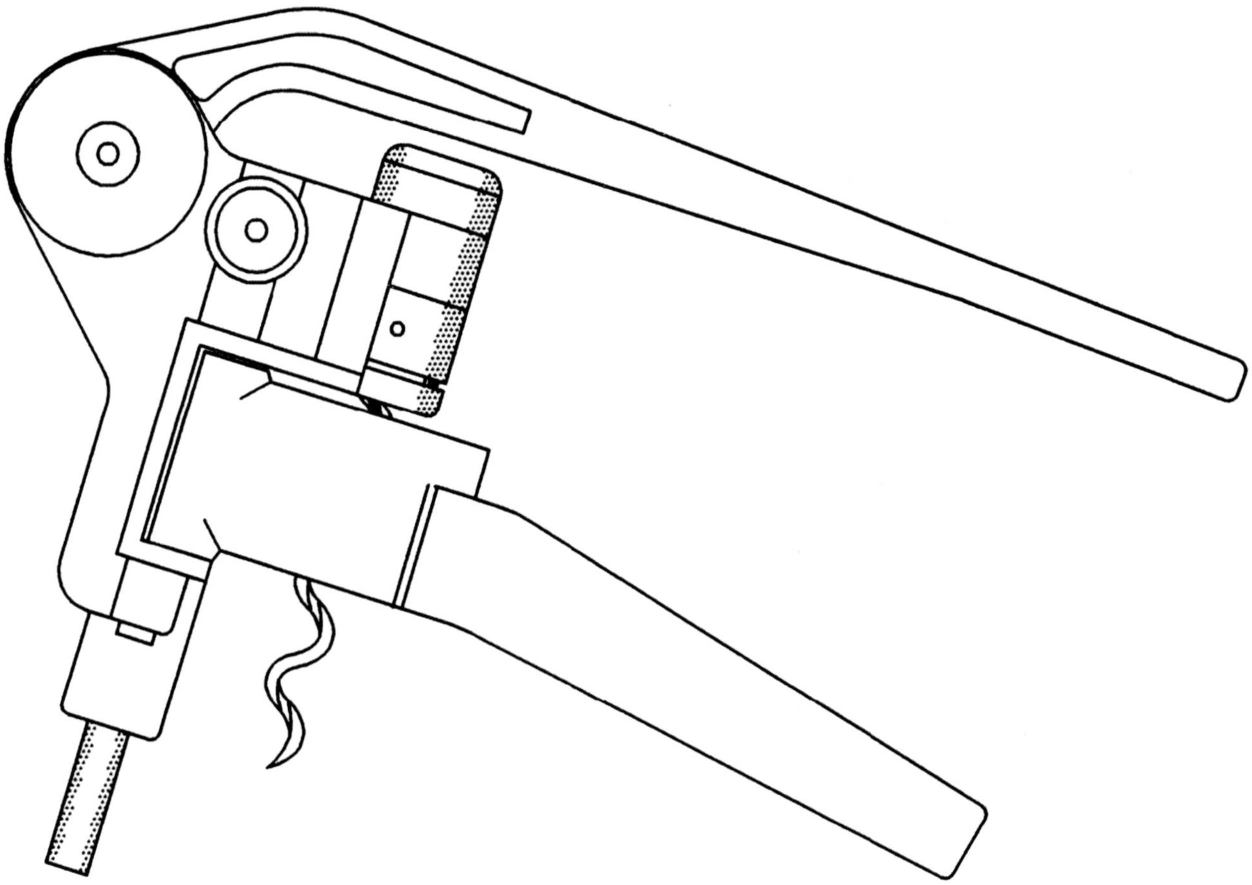

Chapter 6: Leverpull Production Models 1981–1991

The Hallen Company

Although this chapter is "Leverpull Production Models," all of the Screwpull Lever Model corkscrews are called Leverpull. "Leverpull" is the term we used from the very beginning, talking with Herb Allen. The "Leverpull" now consists of several models. "Screwpull" is the registered trademark. Now that there are different models, they are referred to as Screwpull Lever Model, Screwpull Professional Model, Screwpull Professional Model LX, Screwpull 2000 Professional Model, Screwpull Professional Stealth Model, Screwpull Professional Elegance Model, LM 400, LG 410, etc. In general, I will be referring to all of the Screwpull lever models as "Leverpull," also a registered trademark.

The Hallen Company, headquartered in Houston, Texas, manufactured the Leverpull and most all of the Screwpull products until 1991. Herb Allen died on June 12, 1990. Mr. Allen's will provided for the ownership of the Hallen Company (Hallen International) to transfer to Rice University in Houston, Texas. Rice University, not being in the manufacturing business, put the Hallen Company up for auction. LeCreuset, headquartered in France, entered the winning bid. LeCreuset is a kitchenware products manufacturer. From 1991 to the present, Le Creuset has manufactured the Screwpull products. Most of their products started out being made in France, then most of the Lever Model corkscrews were manufactured in Ireland, then India, and, finally, China.

The Leverpulls have a serial numbering system. The Hallen Company started hand-stamping the numbers behind the sliding rack and/or on the inside of one of the bottle-grip handles. These hand-stamped units are the same ones with the Screwpull/Leverpull logos. Being a cumbersome operation in the assembly of the corkscrew, the numbers were changed to machine stamping on the body of the corkscrew, behind the sliding rack. By pulling the pull lever to the two-thirds open position, the rack slides up exposing the number on the body. If you pull the lever to the full open position, the pull lever will obscure the number. Refer to illustrations on pages 82–83. Most examples have a serial number, except the Tortoise Shell version of the Lever Model. For some unknown reason, there is not a number or a "Made in" location stamped or cast onto the inside of the bottle-grip handles. Some later versions of the Lever Models (after "Made in France") do not have "Made in" locations stamped or cast anywhere on the corkscrew. Information on the packing box will tell where it was made. The LM300 was the last Leverpull model that had serial numbers.

Removing a Cork with Leverpull

①

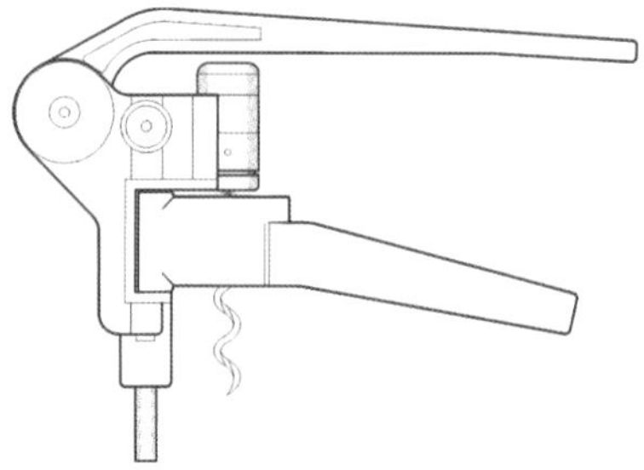

Pick up Leverpull with left hand by grip handles in the closed position.

④

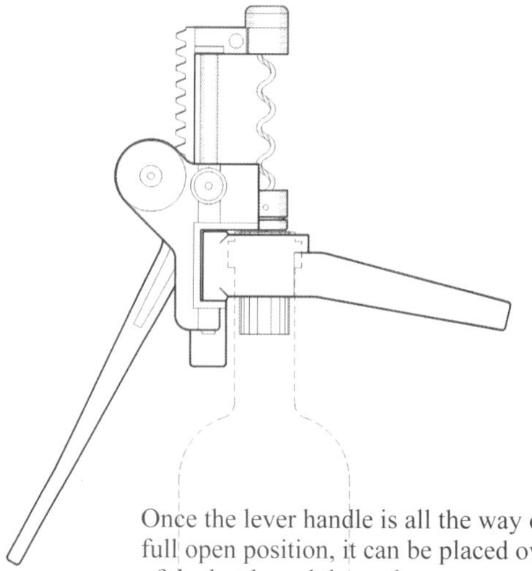

Once the lever handle is all the way over, in the full open position, it can be placed over the top of the bottle and the cork.

②

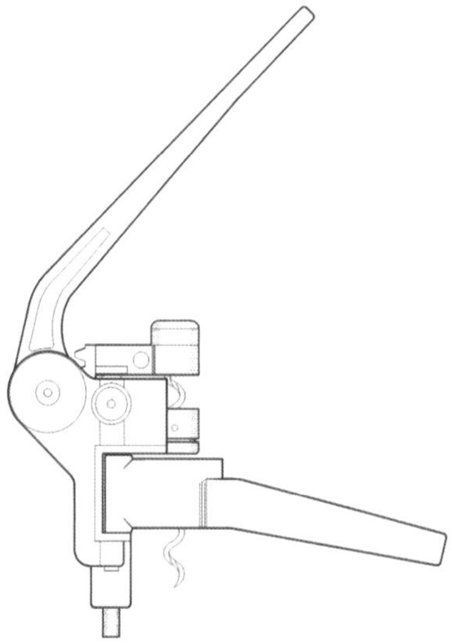

Start pullling up on the lever handle.

⑤

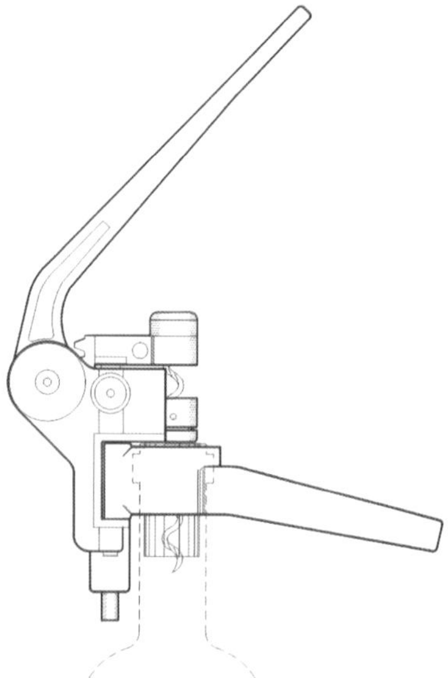

Firmly grip the bottle-grip handles until you see the latch pins protruding from the sides. Slowly pull the lever handle up and over the top until it is in the fully closed position.

③

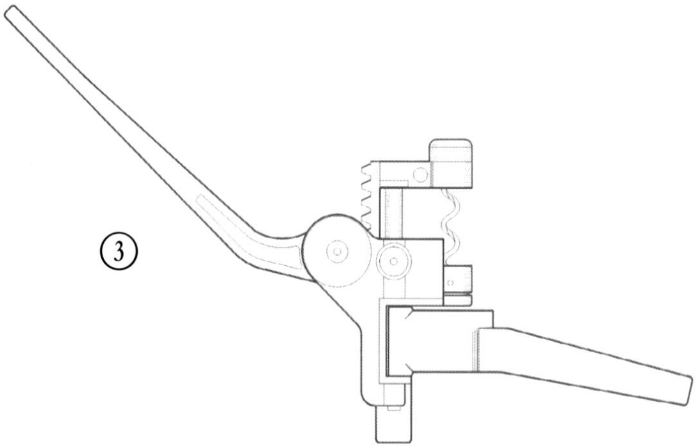

Keep pulling the lever handle up and over the top.

⑥

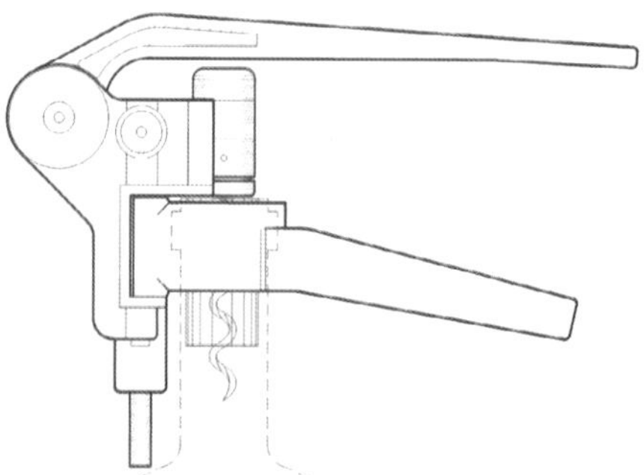

Once you have the lever back into the fully closed position, the screw has fully penetrated the cork.

⑦

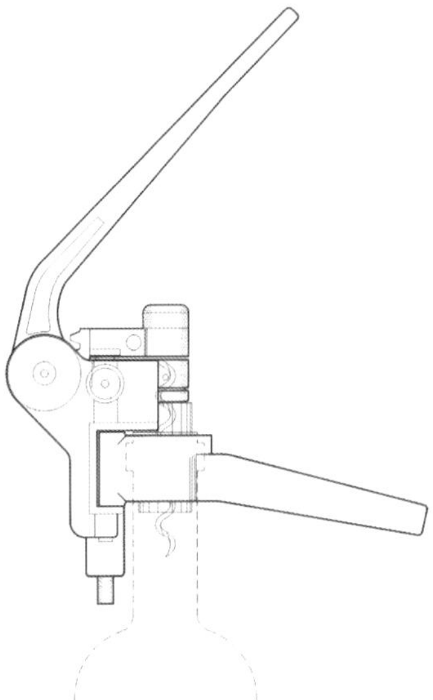

Once the pull lever is in the full open position, the cork has been removed from the bottle.

⑧

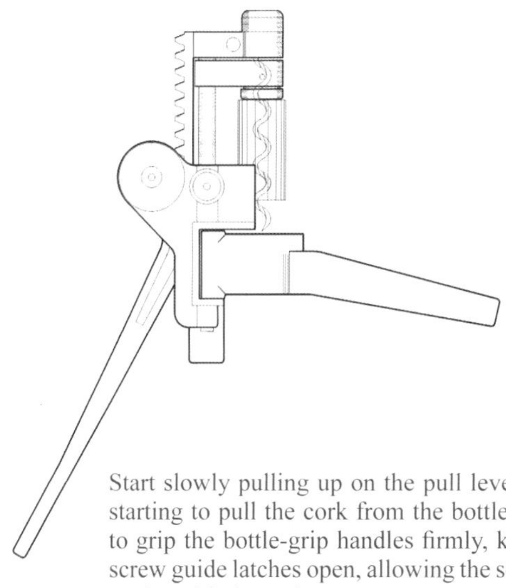

Start slowly pulling up on the pull lever which is starting to pull the cork from the bottle. Continue to grip the bottle-grip handles firmly, keeping the screw guide latches open, allowing the screw guide to rise with the cork.

Removing the Cork from the Screw

①

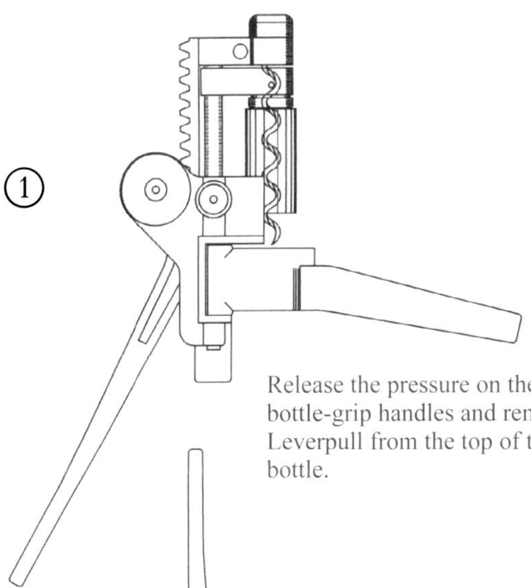

Release the pressure on the bottle-grip handles and remove Leverpull from the top of the bottle.

②

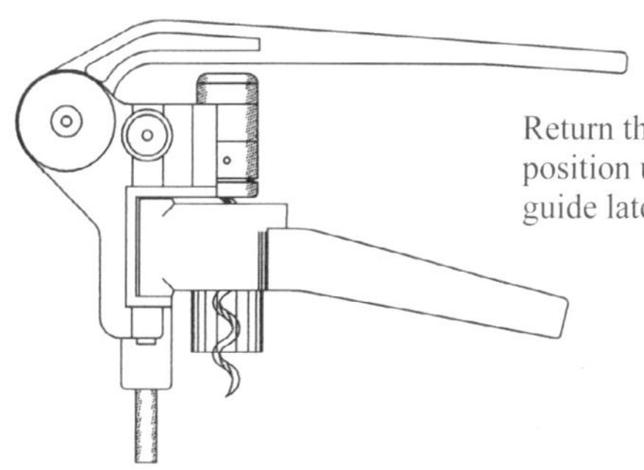

Return the pull lever to the closed position until you hear the screw guide latches snap into place.

③

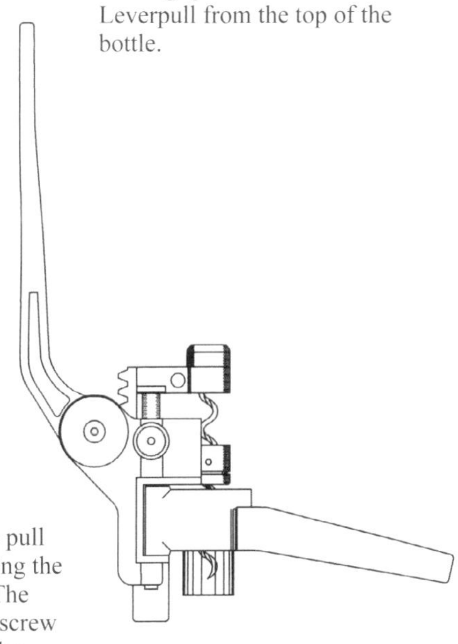

Pull up and over on the pull lever, which starts pulling the screw out of the cork. The latches are holding the screw guide down, allowing the screw to spiral out of the cork.

④

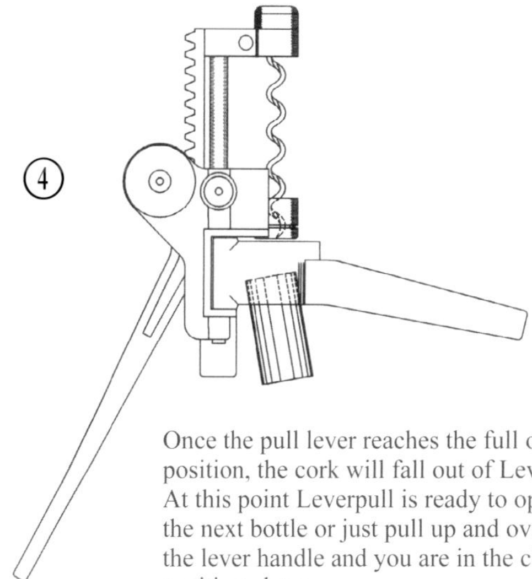

Once the pull lever reaches the full open position, the cork will fall out of Leverpull. At this point Leverpull is ready to open the next bottle or just pull up and over on the lever handle and you are in the closed position, done.

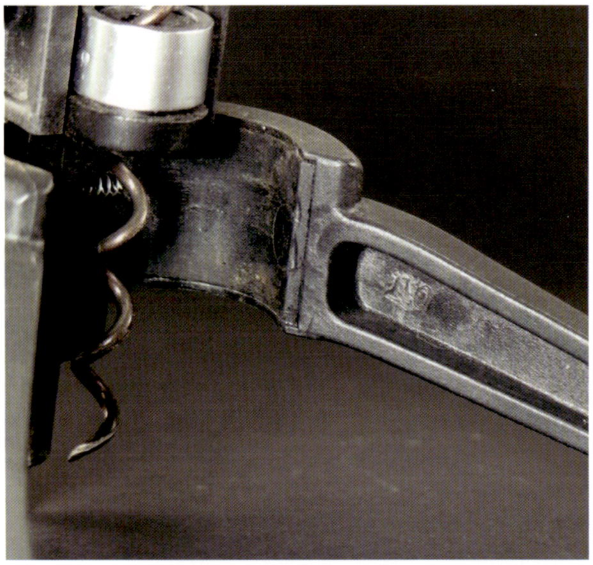

Early Leverpull Production Model #10, made in England, with "LEVER PULL, Patened" under Screwpull in logo area instead of "PATENTED, Pat Pending."

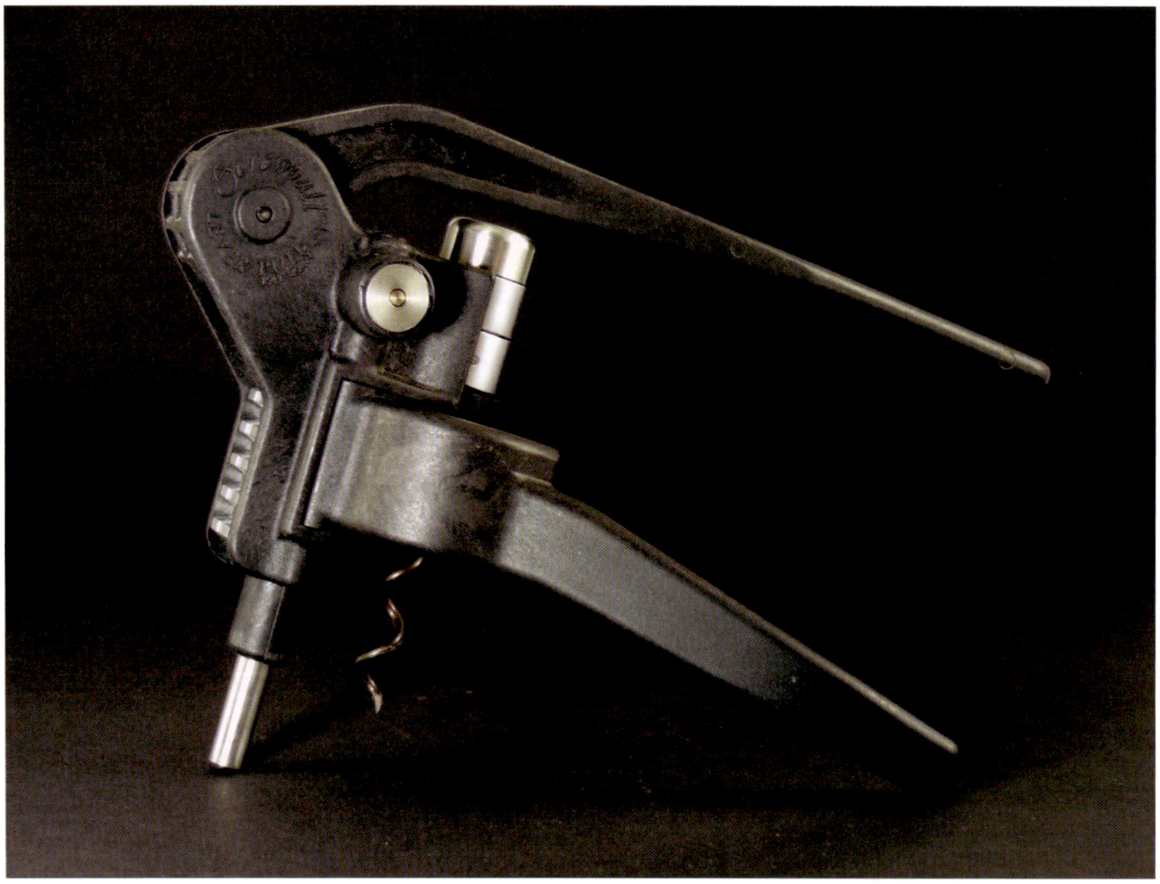

Close-up of logo area.

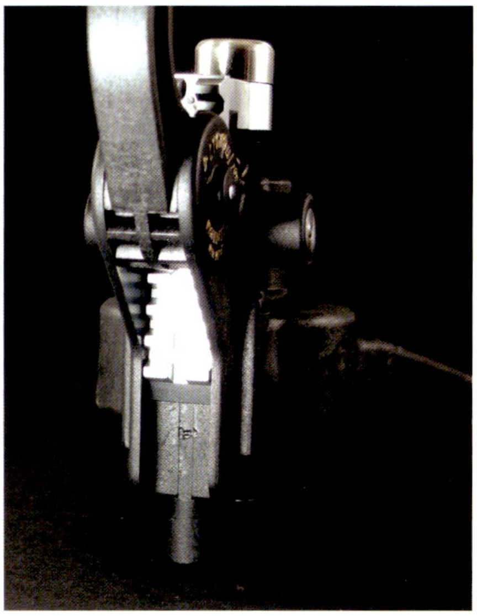

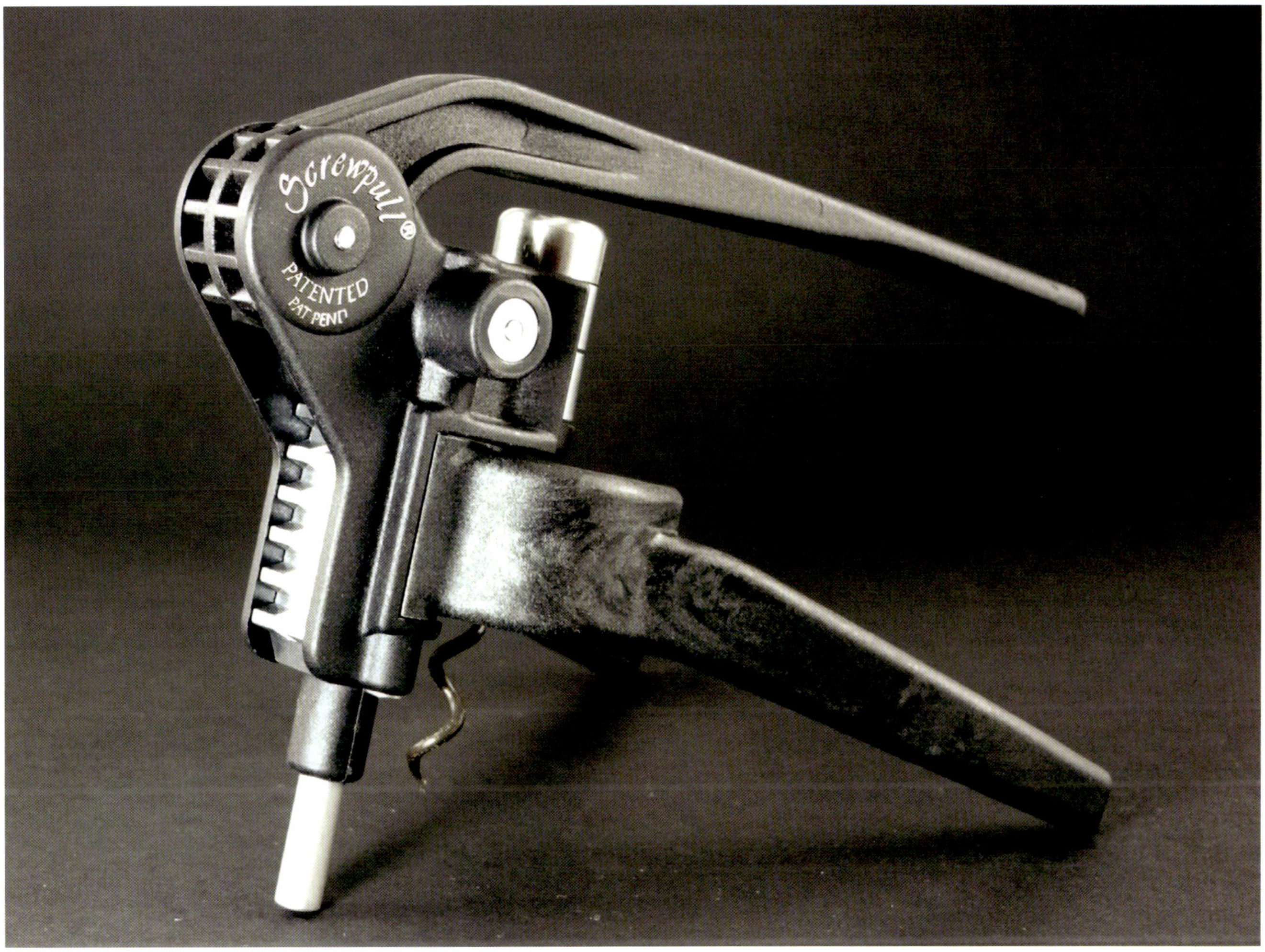

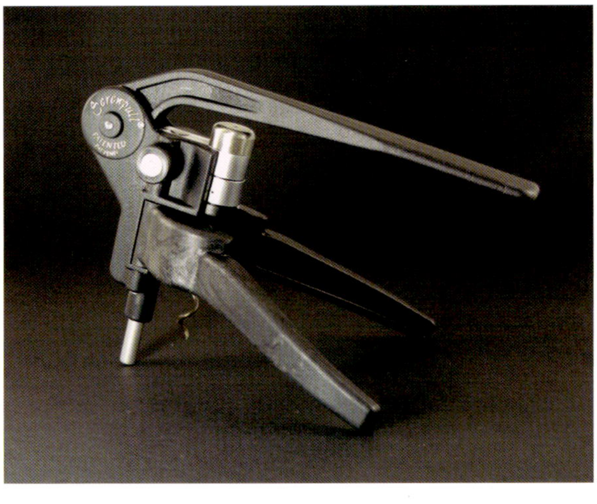

Leverpull Production Model #1, made in England. *Courtesy Loretta Haren.*

Leverpull Production Model #54, made in USA. *Courtesy Loretta Haren.*

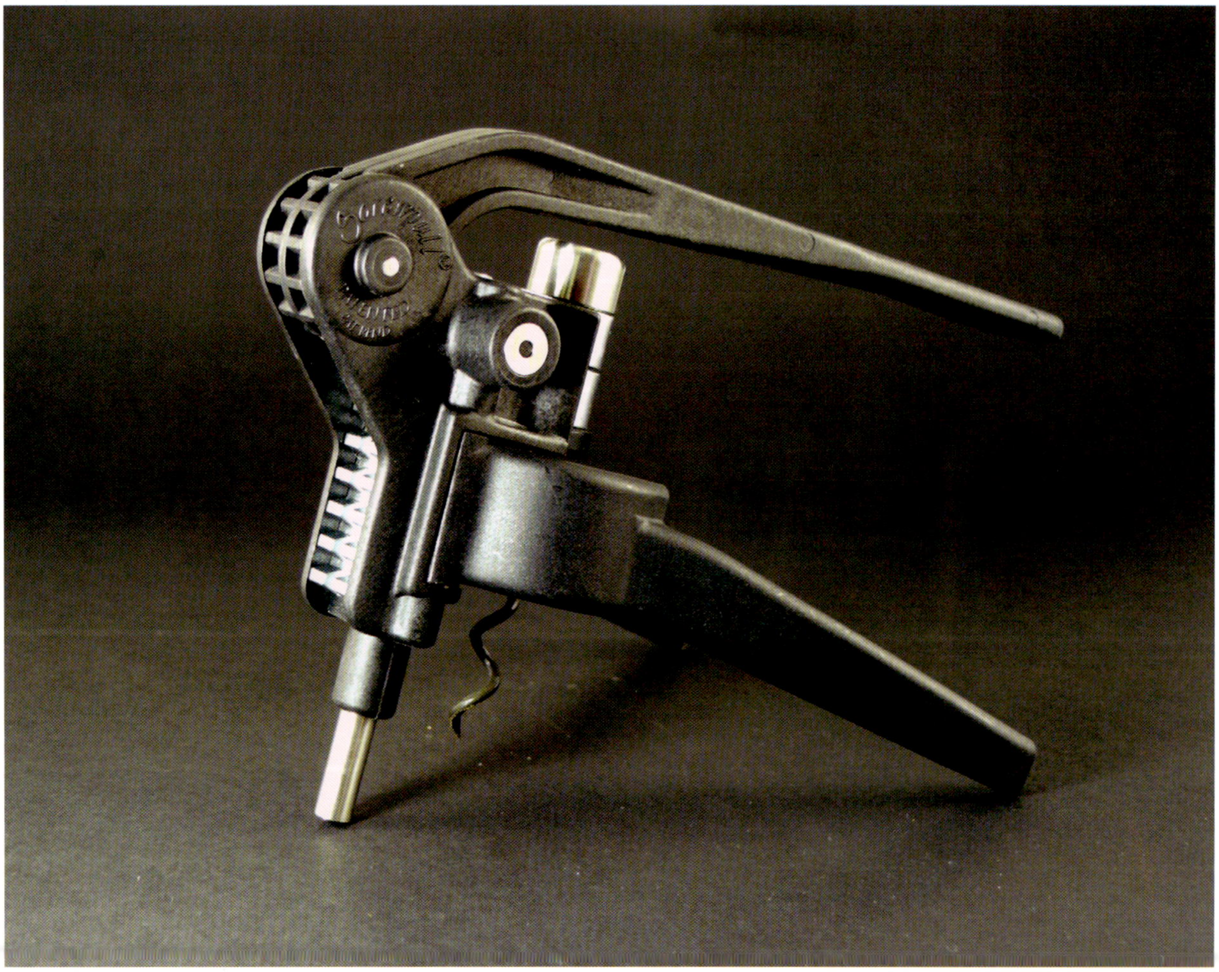

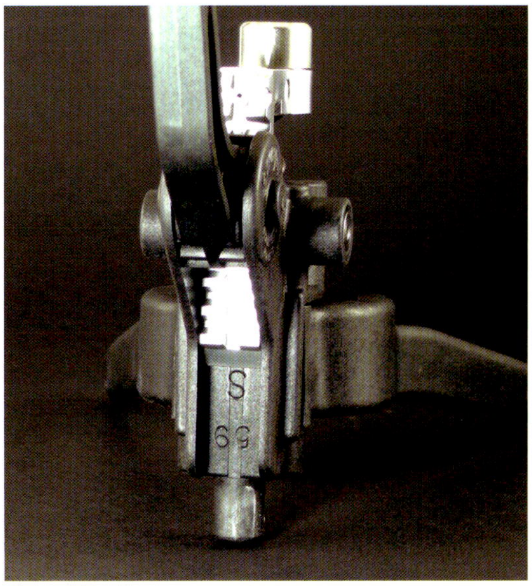

Leverpull Production Model #59-S, made in USA.

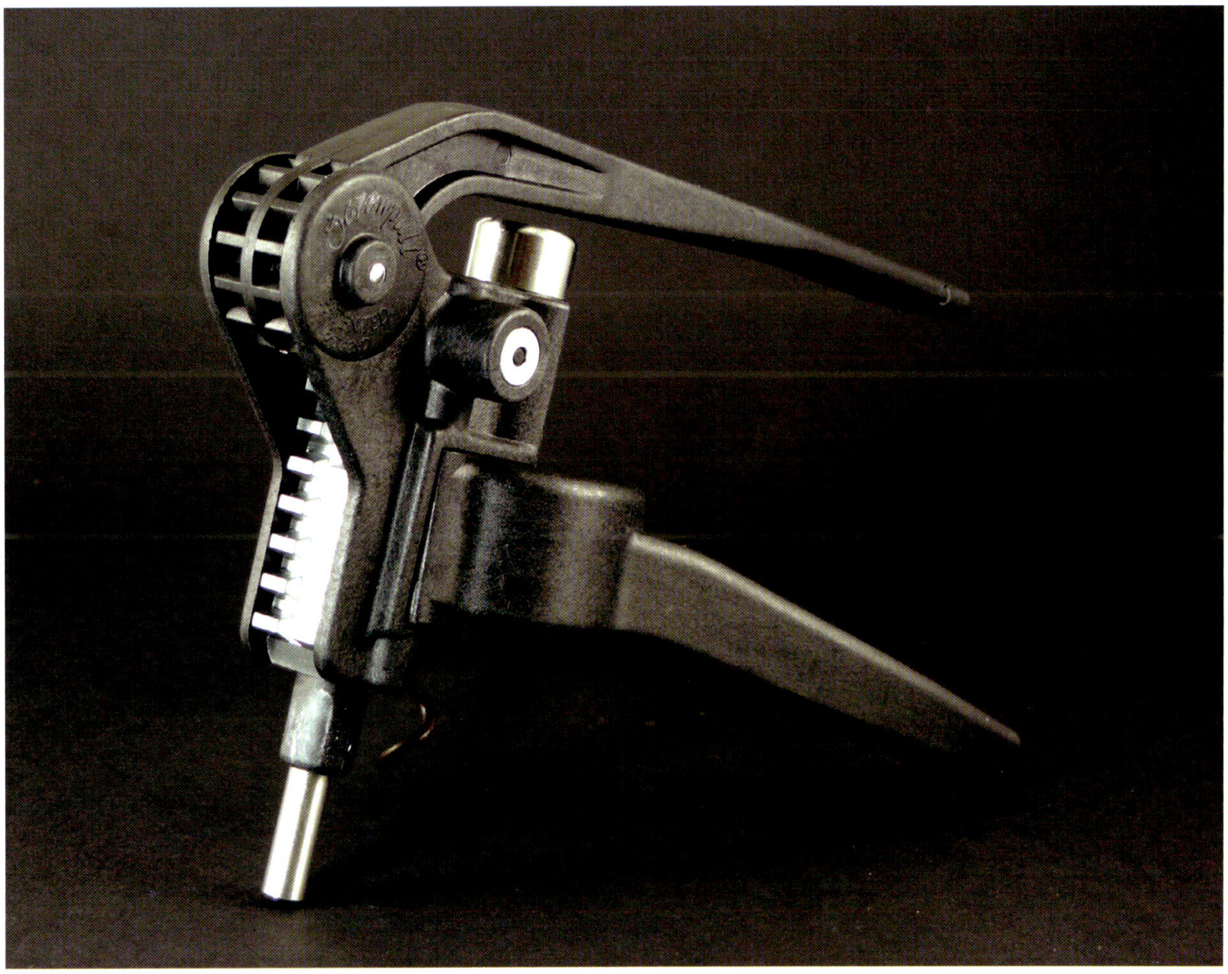

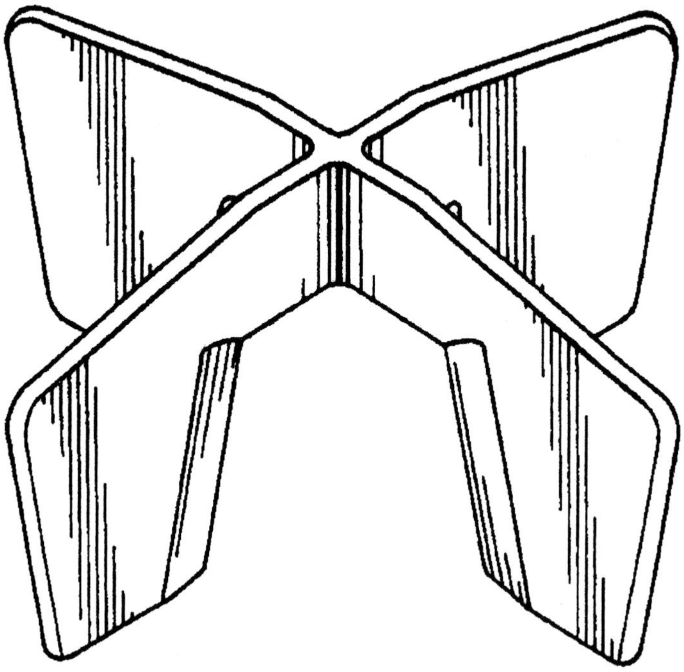

Chapter 7: Screwpull Champagne Star

The Screwpull Champagne Star, Patent No. 4,800,783, January 31, 1989, was developed by Herb Allen to help assist in removing a champagne cork without disastrous results. Herb knew exactly what he wanted to do, so he just started making prototypes from different materials, such as brass, aluminum, and composite plastic resins, to see what would work the best. The Champagne Star evolved rather quickly and went to market in several different packaging styles.

Champagne Star Prototype, composite plastic material.

Champagne Star Prototype, stainless steel.

Champagne Star Prototype, bronze alloy.

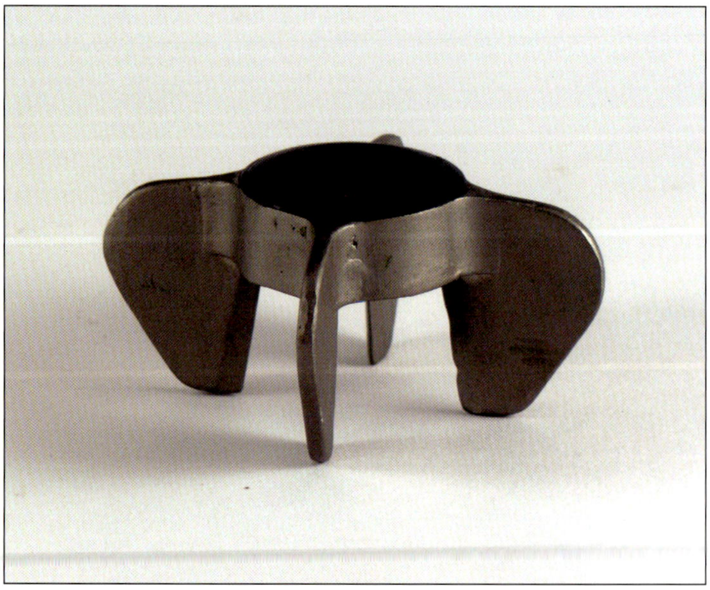

Champagne Star Prototype, rustless metal material.
Courtesy of Loretta Haren.

Champagne Star Prototype, brass.

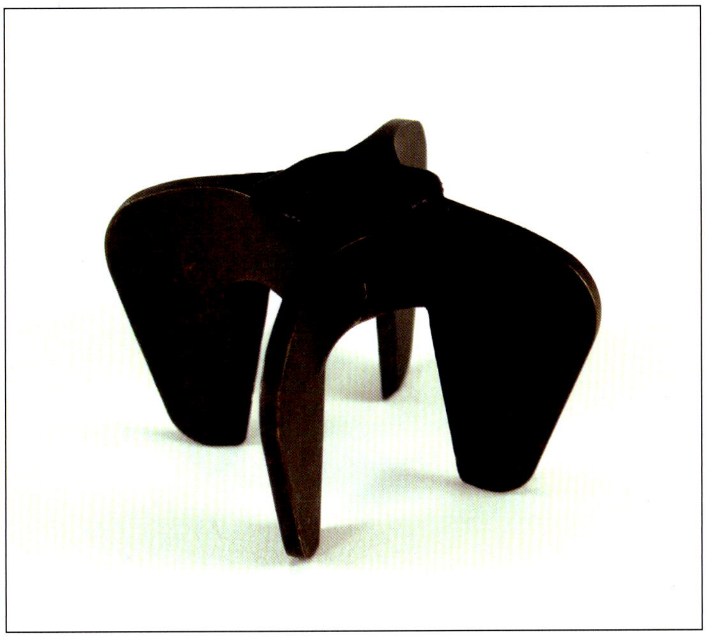

Champagne Star Prototype, cast metal alloy.

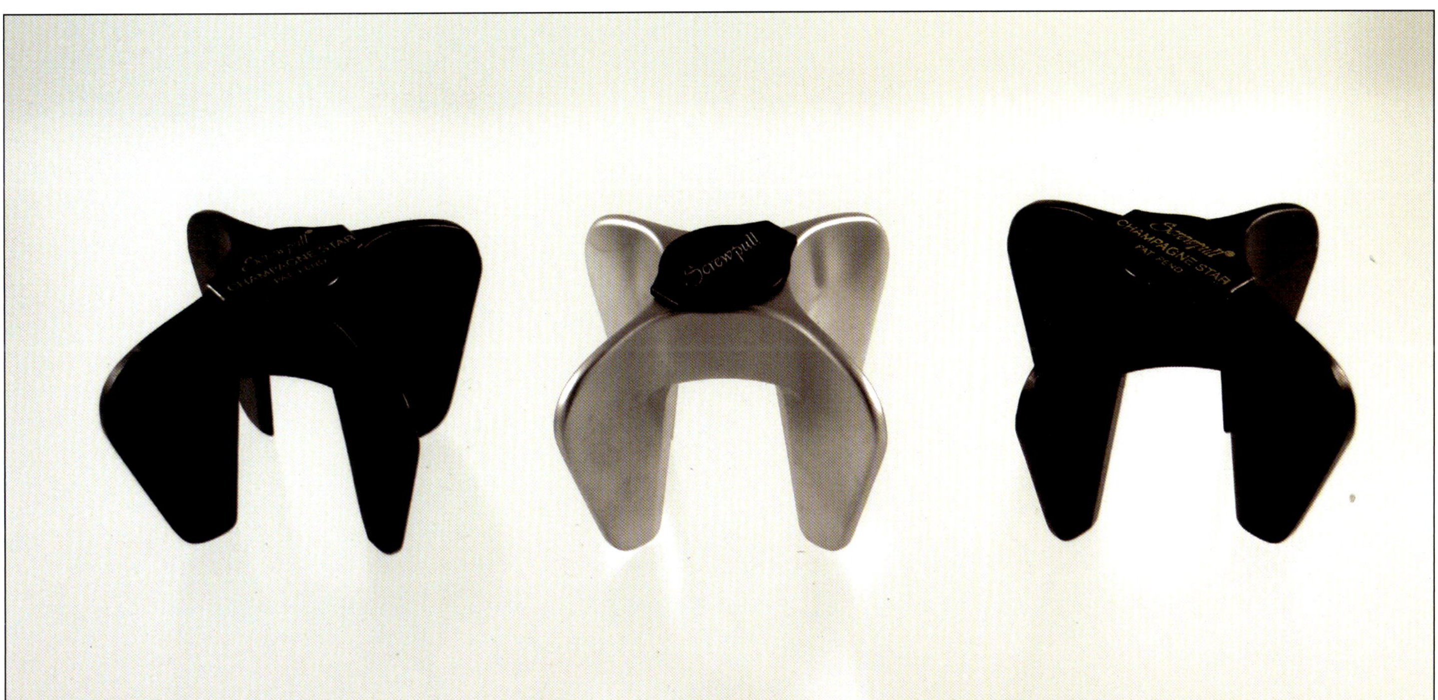

Champagne Star Production Models in black and silver colors.

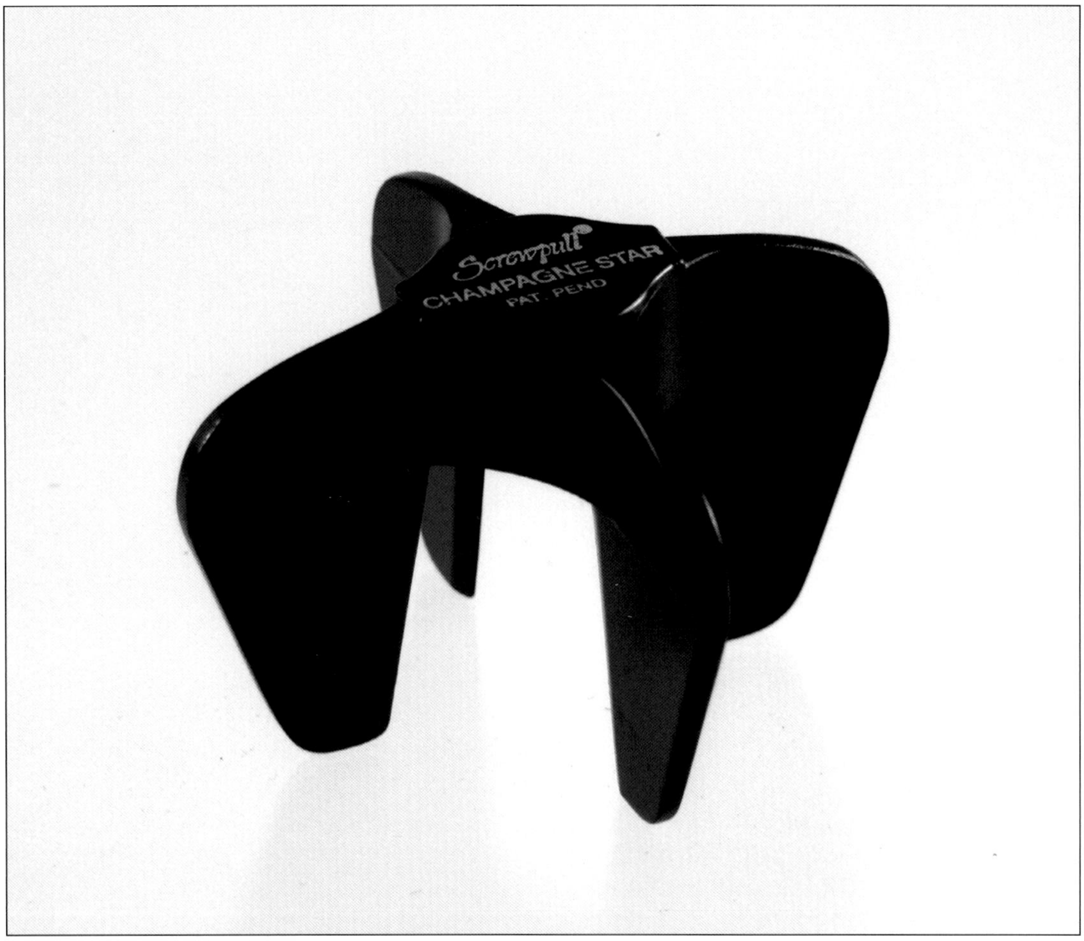

Black Production Model.

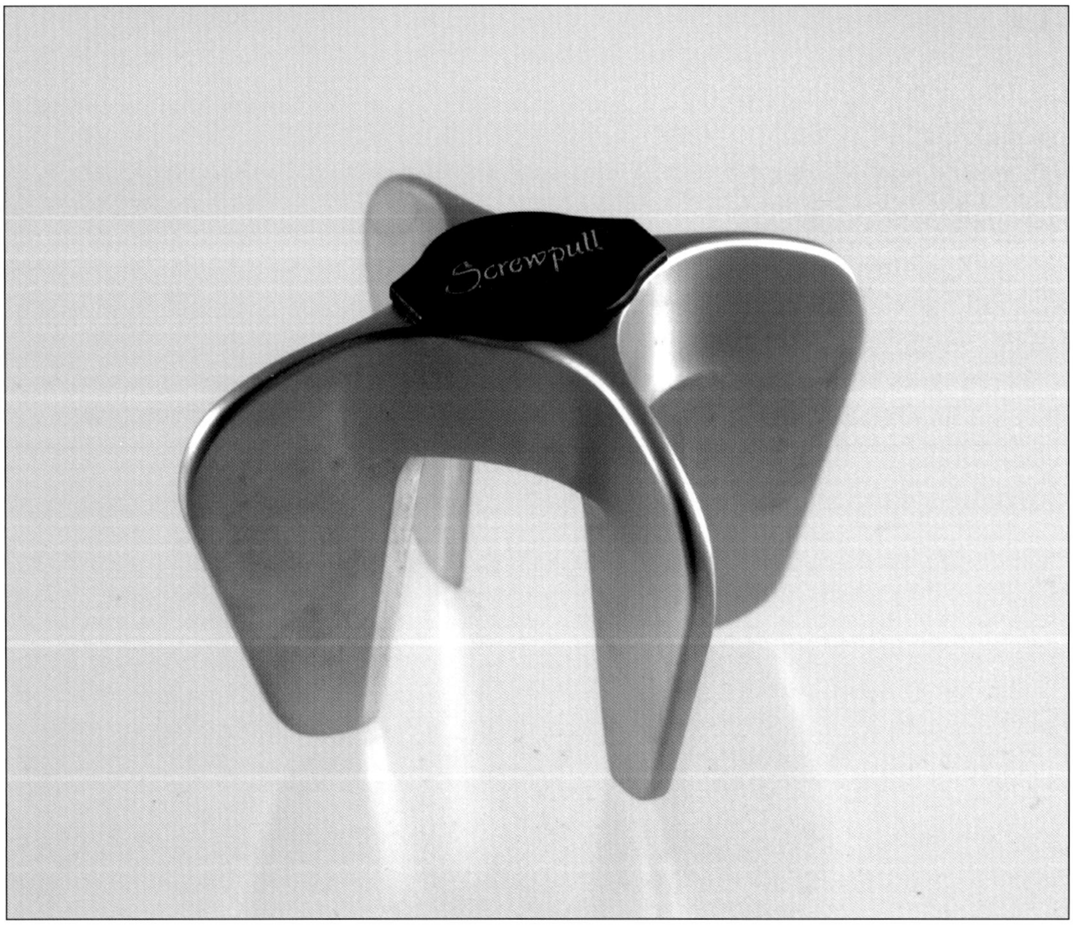

Silver Production Model.

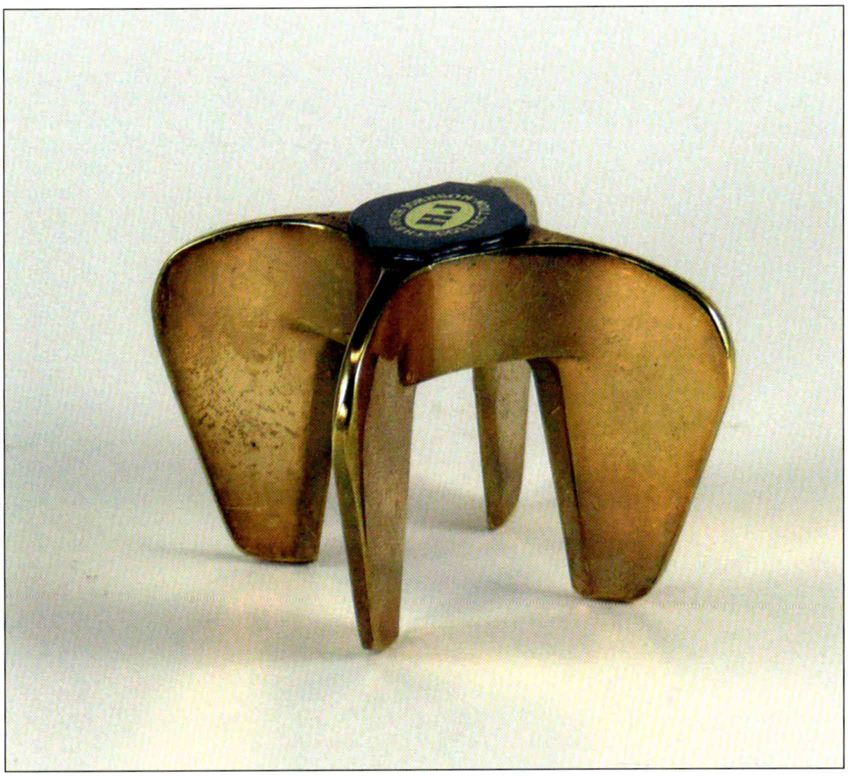

Gold-colored special order for Hugh Johnson sets.

Silver-colored Production Model Champagne Star in box with each half of cover sliding out to access the Star.

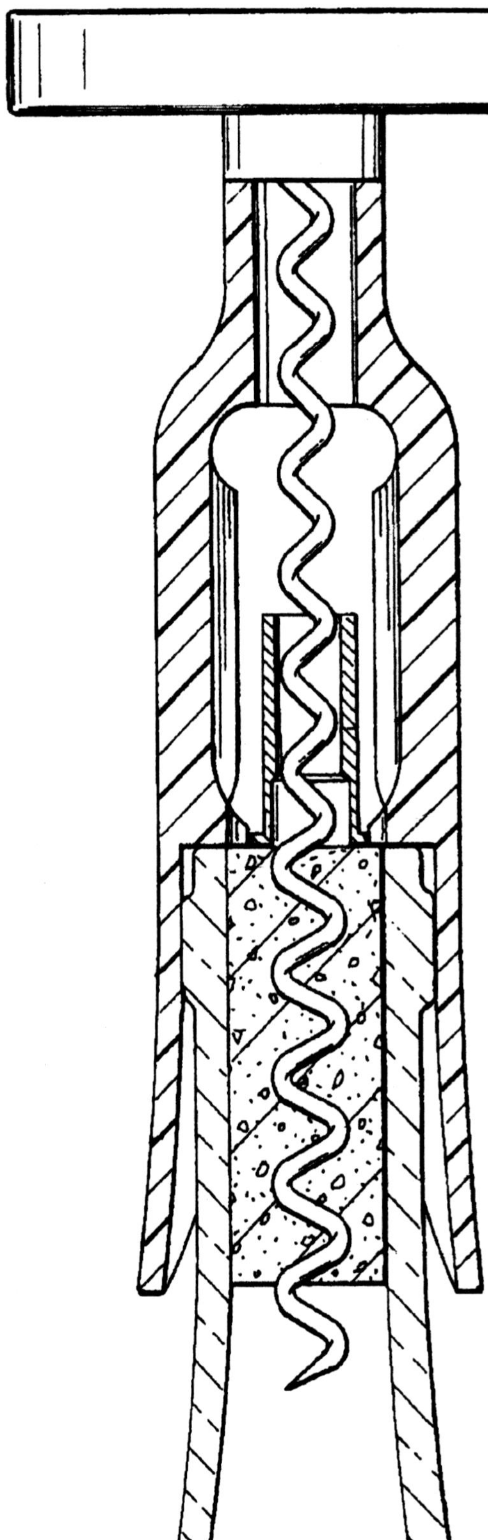

Chapter 8: The Cork Ejector

U.S. Patent Number 4,800,784 states that this is an apparatus for removing corks from bottles and is adapted to allow for automatic ejection of the cork from the apparatus after the cork has been withdrawn from the bottle, by reverse rotation of the corkscrew.

Removing the cork from the simple, basic Screwpull is achieved by reversing the corkscrew handle while the ridges on the frame keep the cork from turning. About halfway down, you then push the handle into the frame, pushing the cork down, past the ridges, where you then hold the cork with your fingers and finish turning the screw in reverse, until the cork turns loose of the screw.

The cork ejector was comprised of a stainless steel tubular nut that would ride up and down the screw to eject the cork. First, you had to devise a way that the stainless steel nut would not release itself from the screw and then be lost in short order. This was achieved by having two different outside diameters of the screw. For about a third of the way from the bottom of the screw, the outside diameter of the screw would be the normal 3/8", but the upper two thirds of the length of the screw would be, say, two thousands of an inch smaller in diameter. The inside diameter of the stainless steel tubular nut would be the same diameter as the upper part of the screw, so that when the nut would reach the larger diameter of the lower part of the screw, it could go no farther, meaning it could not come off the screw and get lost.

Second, the upper bushing hole in the frame had to be the exact diameter as the outside diameter of the stainless steel cork ejector nut, meaning that it would be larger than the screw and is a sloppy fit for the upper part of the screw. This was unacceptable for the engineer. Having two different sized bushing holes in the top of the frame, meaning a newly designed bulkier frame, could solve this, but would make for an even more complex system than we already have. Note in the top right photograph how the handle on the right Screwpull is off-center to the frame because of the sloppy bushing hole.

This system worked if the operator knew exactly what he was doing and how and where to place the nut before starting. It was not people-proof. The extra costs of manufacturing the stainless steel nut and its assembly, and the fact that it was not people-proof, were just not worth it. The simple Screwpull has its screw and handle attached in one operation and then the screw inserted into the bushing hole and then into the box. That was it. This system would require the stainless steel ejector nut to be placed onto the screw from the top, with flanges down, then inserted into the frame up into the bushing hole from the bottom side, and then install the handle onto the screw dangling out of the top of the frame, all causing assembly problems.

For obvious reasons the Cork Ejector Screwpull was never manufactured, except for these two prototypes and the Patent Prototype. Later, LeCreuset incorporated a similar cork ejector system with the Activ Red Ball Triology Model.

Screwpull Cork Ejector device on Version 3 Table Model with a separate cork ejector tube.

Close-up of ejector tube.

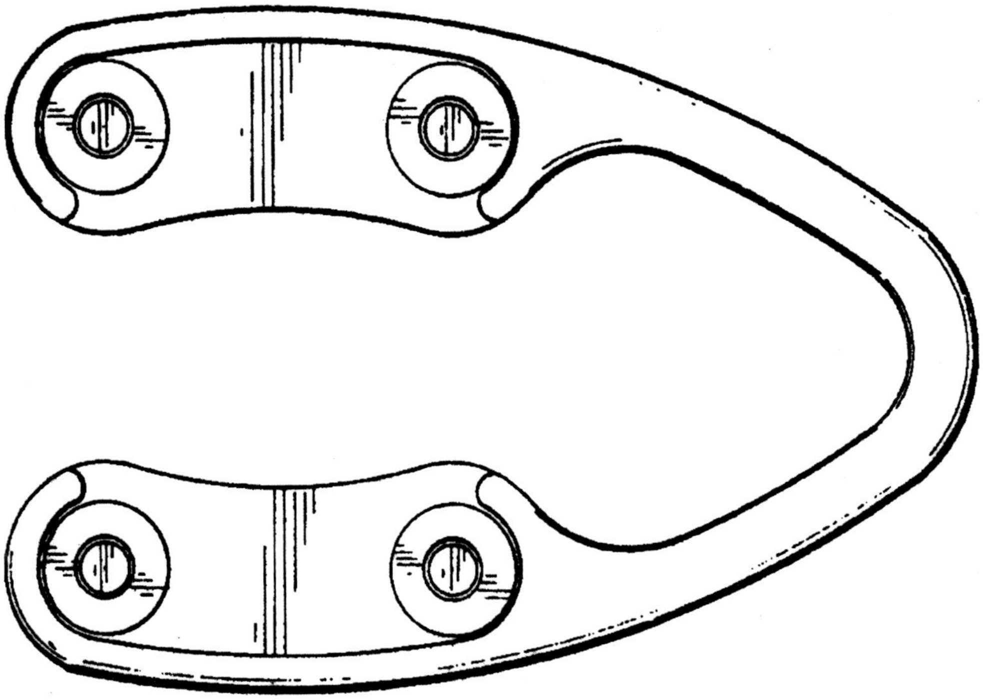

Chapter 9: The Screwpull Foil Cutter

The Screwpull foil cutter, Patent No. 4,845,844, July 9, 1989, and its Design Patent No. Des. 297,201, August 14, 1988, was first marketed as a stand-alone item, packaged in a small box. It was later included with the Table Model as a set. The foil cutter rested in a notch on the stand of the Table Model corkscrew, where it had less chance of getting lost in a drawer or on the bar top. It had a place with the corkscrew. It also was sold as a set with the Pocket Model Screwpull and the Leverpulls.

The Foil Cutter combined as a set with the Table Model Prototype. Table Model, Foil Cutter, and stand on left are Production Models. The stand on the right is a Prototype.

The Foil Cutter combined as a bright red set with the Version 3 Table Model Production Model.

Green Hugh Johnson-printed Foil Cutter from a Spinhandle Pocket Model set.

A few foil cutter knockoffs—the original Screwpull Foil Cutter is at top-center.

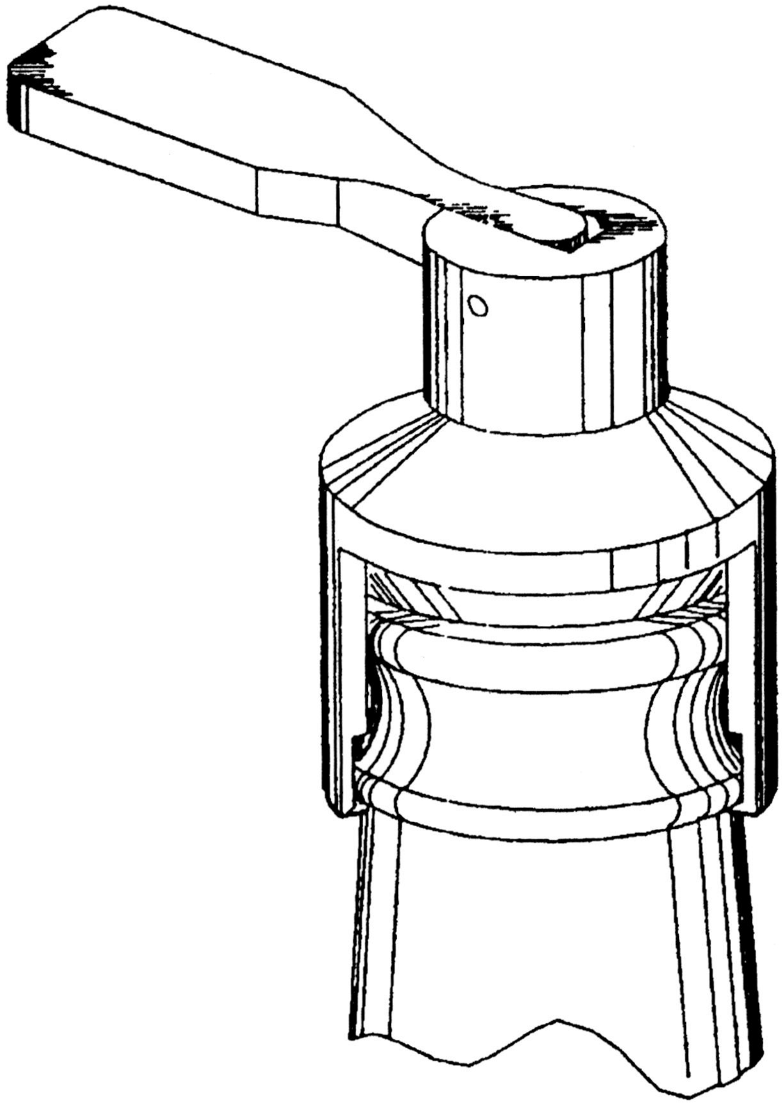

Chapter 10: The Champagne Seal

The Champagne Seal, Patent No. 5,056,676, was developed by Herb Allen to seal an unfinished bottle of champagne. Once you opened a bottle of champagne and did not finish it, you probably would like to save the remaining nectar until tomorrow without it losing its sparkle and fizz. Install the Screwpull Champagne Seal onto the bottle and it will keep the champagne fresh for several more days.

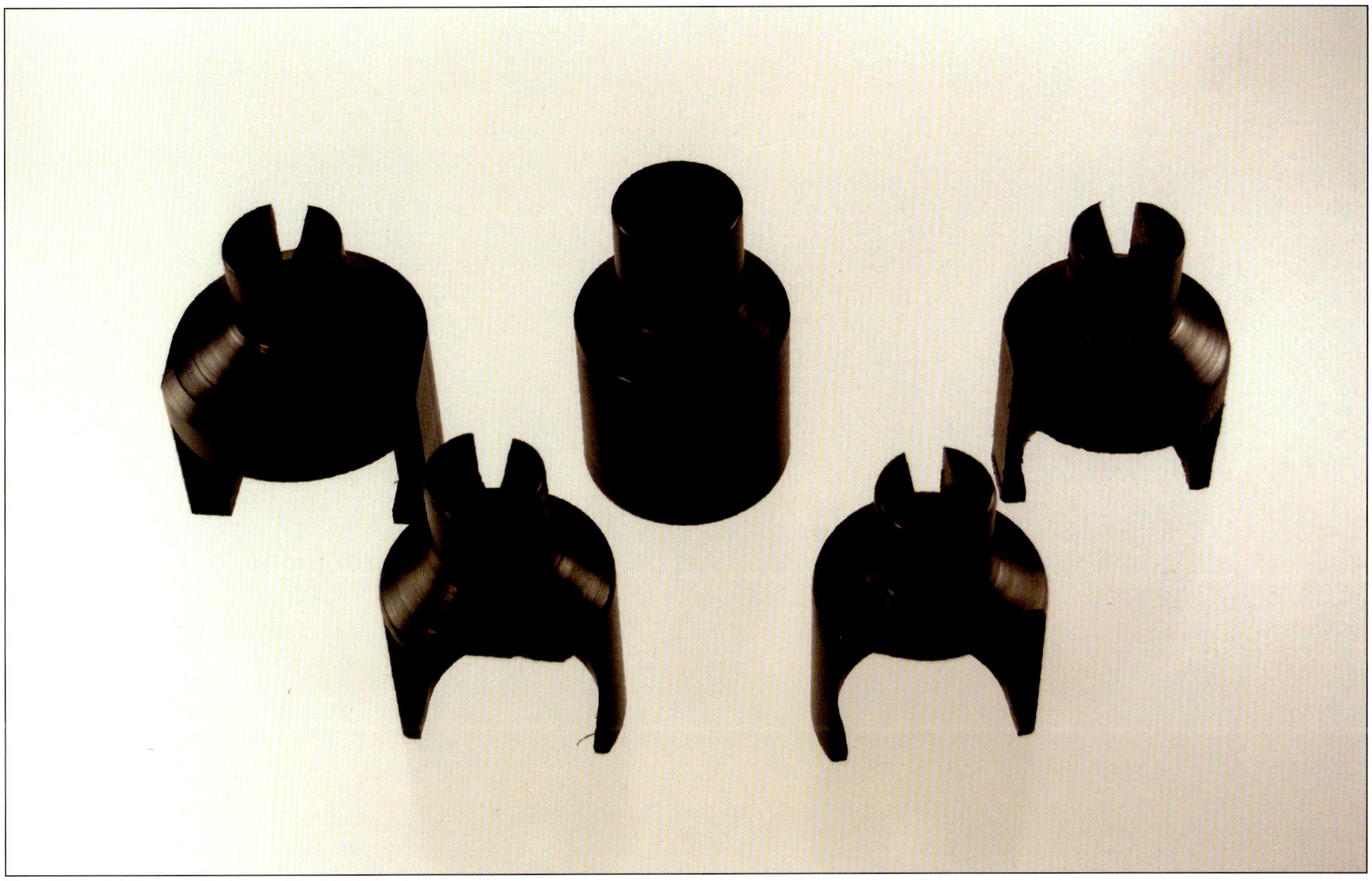

Foreword to Chapter 11

by Margaret Anderson

Herb Allen's patent attorney throughout the latter half of his career in Houston, Texas

During his first career in oil and gas, Herbert Allen's many inventions resulted in hundreds of patents. In retirement, he turned his attention to devices inspired by his outstanding collection of wine, eventually installing machining equipment, a spring winder, and various other tools in his home in order to better conduct experiments.

His first wine-related invention, and the first patented, would eventually be sold as the Screwpull "Lever Model" wine opener. But before the Lever Model could be commercialized, Herb had a brainstorm: the same nonstick coating he had used in the Lever Model could allow him to produce a much simpler device. Thus, the Screwpull Table Model was born.

The Table Model, in turn, led to the Hallen Company for marketing the product, and to a second career for Herb. In time, the Table Model was also patented, as were several improvements and variations, and later, a device for opening champagne and one for cutting and removing foil from the tops of wine bottles. In time, all of these improvements, variations, and other devices were commercialized by Hallen Company, as was the Lever Model.

The Table Model enjoyed particularly great commercial success. Unfortunately, it was so widely copied by foreign manufacturers that it eventually became impractical to take legal action against all patent infringers. Indeed, in some cases, it was difficult or impossible to trace the identity of the actual producer of an infringing product.

Herb took a philosophical view of these infringements, often noting that, "Imitation is the sincerest form of flattery."

Chapter 11: Patent Drawings

U.S. Corkscrew Patents Issued to Herbert Allen of Houston, Texas

Leverpull	U.S. Pat. No. 4,253,351	Filed: Jul 9, 1979 Issued: Mar 3, 1981
Screwpull	U.S. Pat. No. 4,276,789	Filed: Jul 17, 1978 Issued: Jul 7, 1981
Screwpull	U.S. Pat. No. 4,291,597	Filed: Jul 17, 1978 Issued: Sep 29, 1981
Screwpull	Pat. No. B1- 4,291,597 (Re-examination)	Requested: Jun 10, 1982 Certif. Issued: May 28, 1985
Screwpull	U.S. Pat. No. 4,377,096	Filed: Jan 7, 1981 Issued: Mar 22, 1983
Screwpull	U.S. Pat. No. 4,429,444	Filed: Issued: Feb 7, 1984
SpinHandle	U.S. Pat. No. 4,703,673	Filed: Apr 8, 1985 Issued: Nov 3, 1987
Champagne Star	U.S. Pat. No. 4,800,783	Filed: Jan 5, 1987 Issued: Jan 31, 1989
Cork Ejector	U.S. Pat. No. 4,800,784	Filed: Mar 15, 1988 Issued: Jan 31, 1989
Foil Cutter	U.S. Pat. No. 4,845,844	Filed: Aug 24, 1988 Issued: Jul 11, 1989
Champagne Seal	U.S. Pat. No. 5,056,676	Filed: Jan 4, 1991 Issued: Oct 15, 1991

The following pages show the frontal page and the drawing pages of Herb Allen's Screwpull, Leverpull, and related accessory patents.

United States Patent [19]

Allen

[11] **4,253,351**

[45] **Mar. 3, 1981**

[54] **CORK EXTRACTOR**

[75] Inventor: **Herbert Allen,** Houston, Tex.

[73] Assignee: **Hallen Company,** Houston, Tex.

[21] Appl. No.: **56,281**

[22] Filed: **Jul. 9, 1979**

[51] **Int. Cl.**[3] **B67B 7/00**

[52] **U.S. Cl.** .... **81/3.38 A**

[58] **Field of Search** .... 81/3.31, 3.33, 3.37, 81/3.38 R, 3.38 A, 3.48

[56] **References Cited**

U.S. PATENT DOCUMENTS

532,575	1/1895	Morgan	81/3.31
562,645	6/1896	McMaster	81/3.33
593,698	11/1897	Stoll	81/3.37
620,949	3/1899	Morgan	81/3.33
644,088	2/1900	Morgan	81/3.33
676,205	6/1901	Rogginger	81/3.33
678,773	7/1901	Coomber	81/3.31
776,152	11/1904	Strohacker et al.	81/3.33
845,608	2/1907	Bitner	81/3.33

FOREIGN PATENT DOCUMENTS

2340898	9/1977	France	81/3.48

Primary Examiner—James G. Smith

Attorney, Agent, or Firm—Browning, Bushman & Zamecki

[57] **ABSTRACT**

Apparatus for extracting a cork from a bottle comprising a corkscrew rotatably mounted on a carrier, the carrier in turn being mounted for longitudinal reciprocation on a frame. The apparatus further includes a control nut having a screw passage positioned to receive the corkscrew and configured to mate with the configuration of the corkscrew whereby, upon longitudinal movement of the corkscrew in the screw passage, rotational movement will be imparted to the corkscrew. A latch, engageable and releasable independently of the force of gravity is provided for releasably latching the control nut to the frame to restrain relative longitudinal movement therebetween. Relative rotation between the control nut and the frame is also prevented. A bottle-engaging assembly is connected to the frame for positioning a bottle with respect to the frame in longitudinal alignment with the screw passage. The bottle-engaging assembly is associated with the latch and, when a bottle is engaged therein, is operative cooperatively with the engaged bottle to release the latch. The apparatus further comprises guide means laterally spaced from the corkscrew and cooperative between the frame and the carrier for guiding the latter in a longitudinal path. Means are also provided for restricting rotation of the corkscrew as it is raised from the bottle with the cork.

31 Claims, 26 Drawing Figures

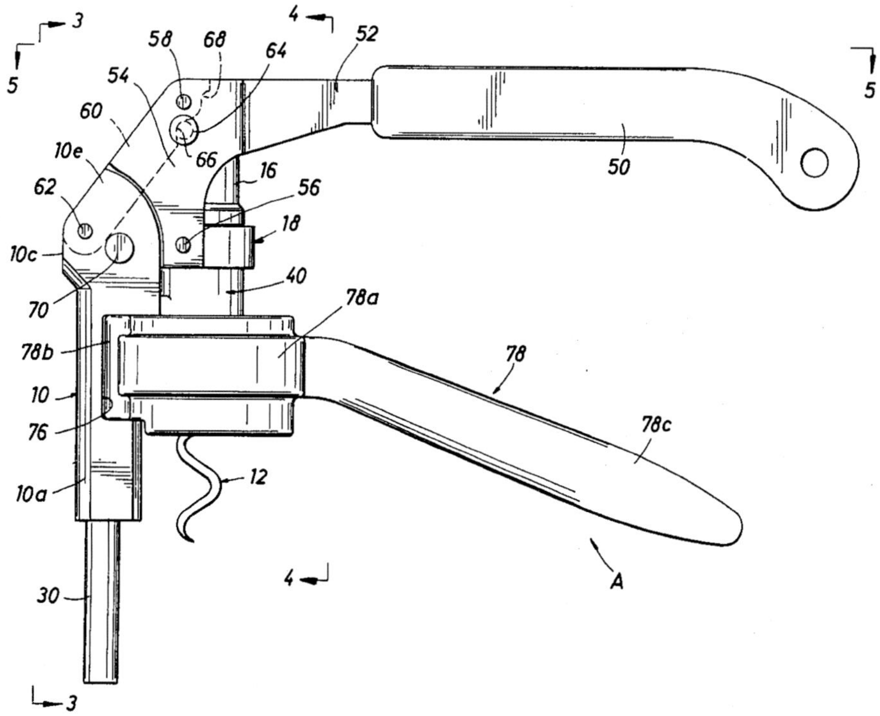

U.S. Patent Mar. 3, 1981 Sheet 1 of 9 4,253,351

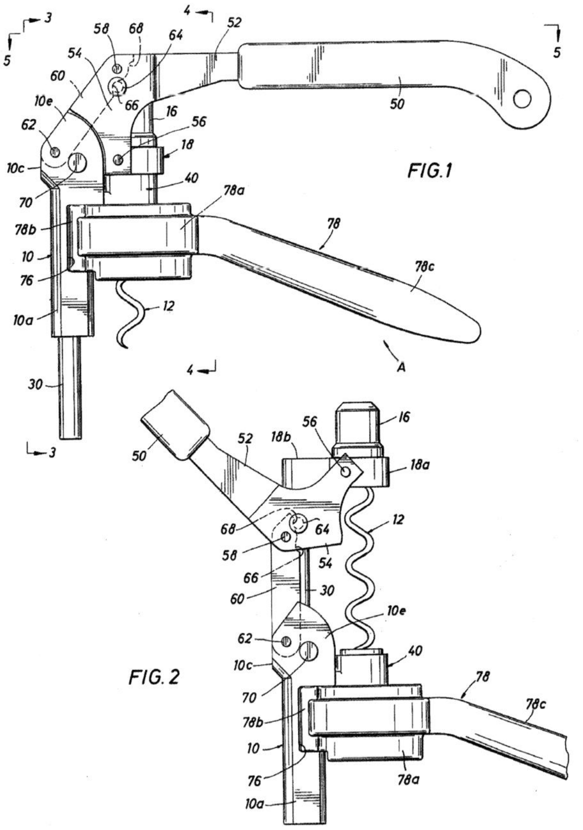

FIG. 1

FIG. 2

U.S. Patent Mar. 3, 1981 Sheet 2 of 9 4,253,351

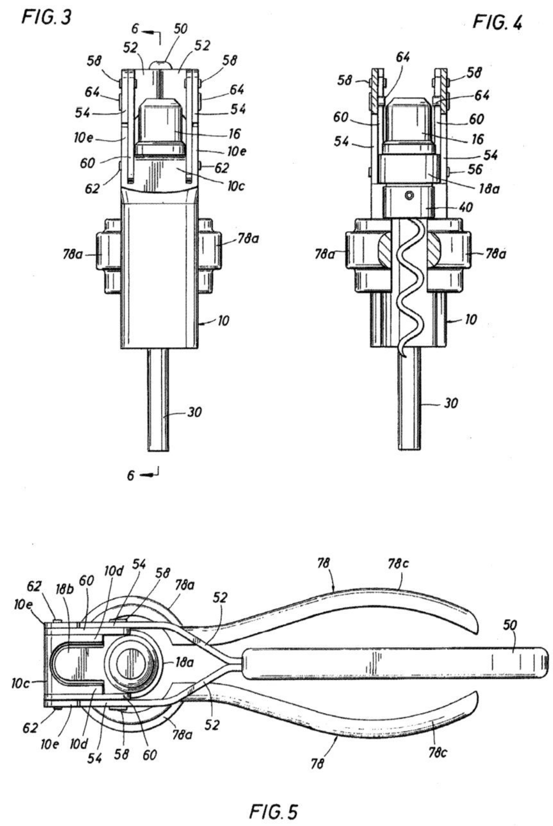

FIG. 3

FIG. 4

FIG. 5

U.S. Patent Mar. 3, 1981 Sheet 3 of 9 4,253,351

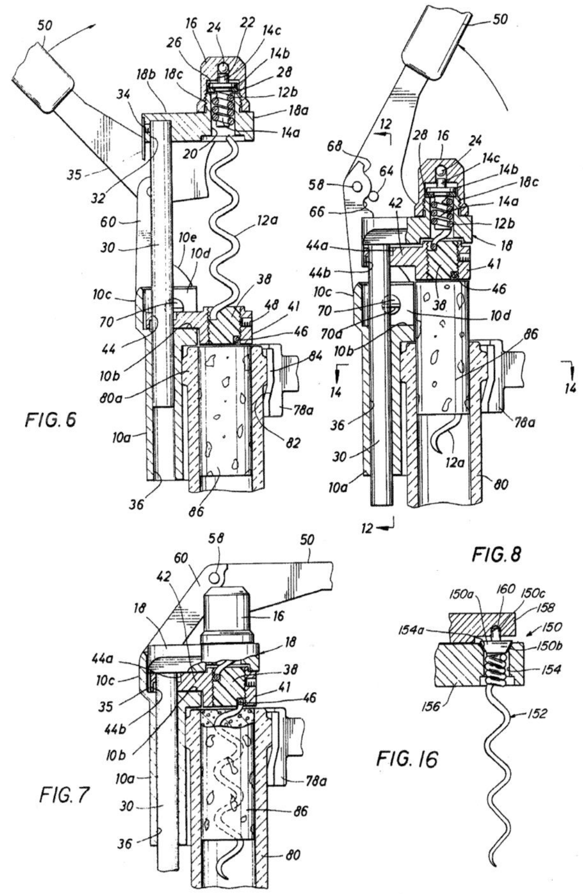

FIG. 6

FIG. 7

FIG. 8

FIG. 16

U.S. Patent Mar. 3, 1981 Sheet 4 of 9 4,253,351

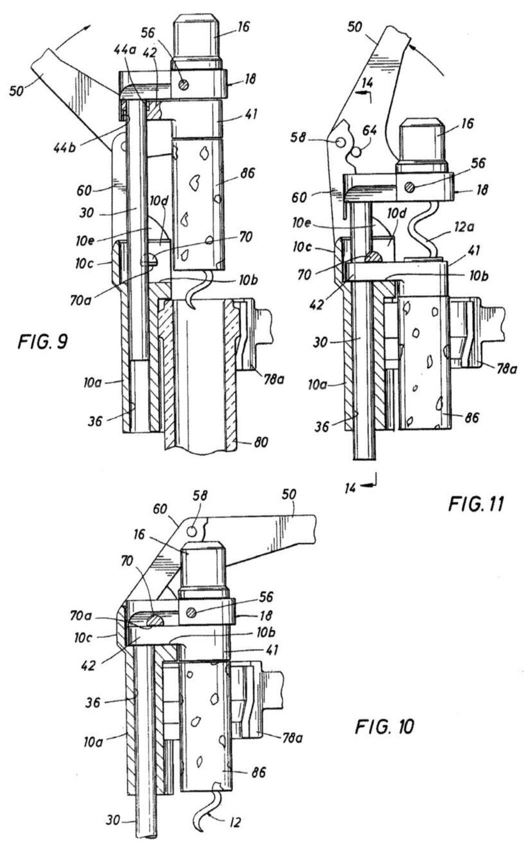

FIG. 9

FIG. 11

FIG. 10

U.S. Patent Mar. 3, 1981 Sheet 5 of 9 4,253,351

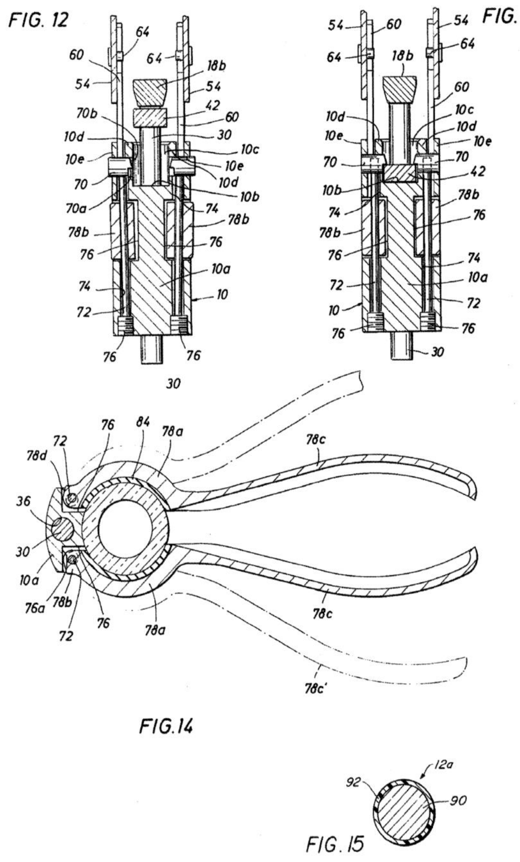

U.S. Patent Mar. 3, 1981 Sheet 6 of 9 4,253,351

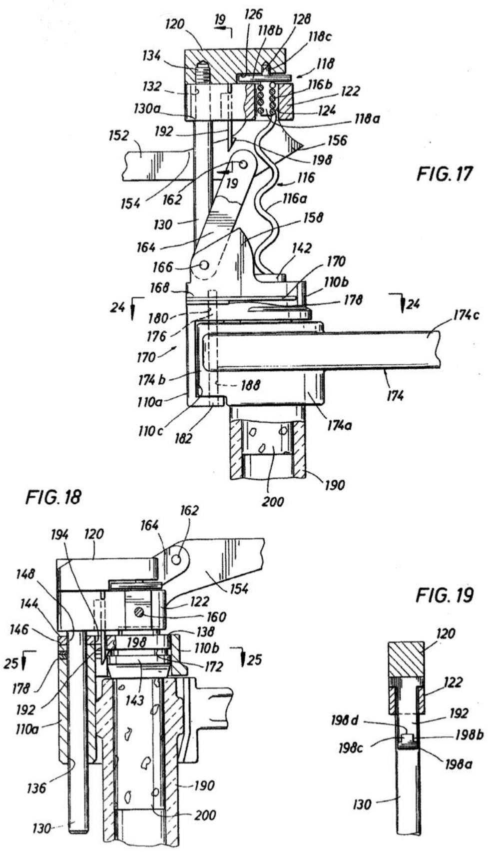

U.S. Patent Mar. 3, 1981 Sheet 7 of 9 4,253,351

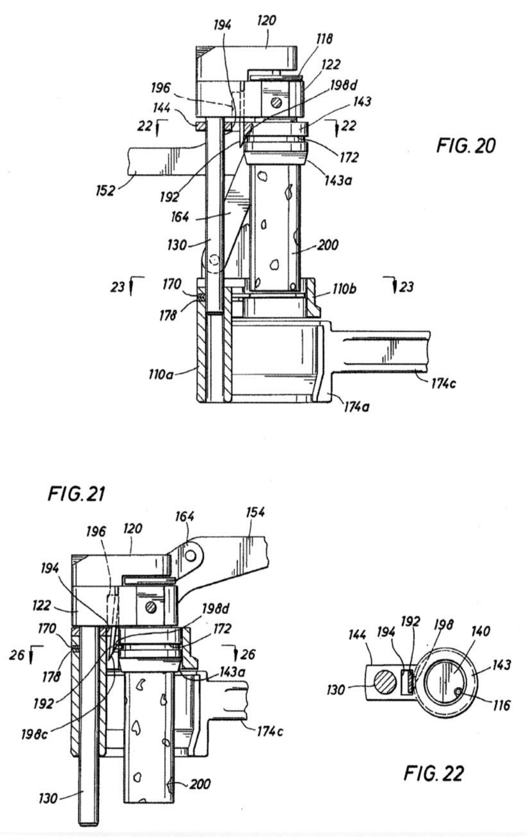

U.S. Patent Mar. 3, 1981 Sheet 8 of 9 4,253,351

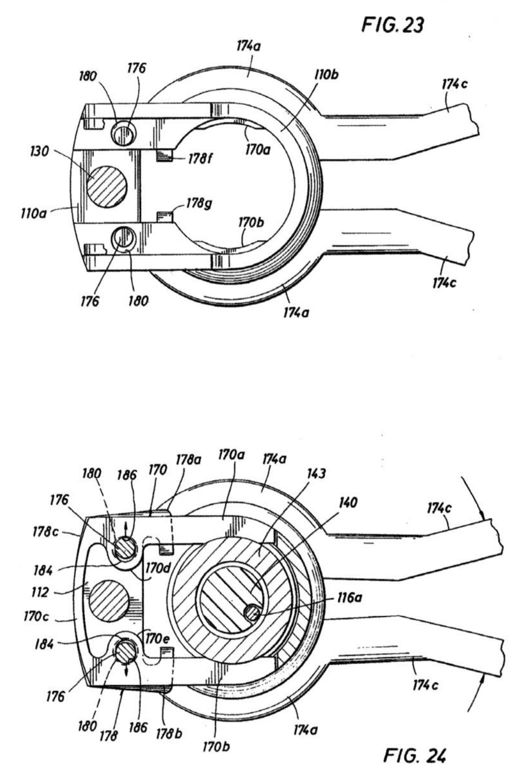

United States Patent [19]

Allen

[11] 4,276,789

[45] Jul. 7, 1981

[54] CORK EXTRACTOR

[75] Inventor: **Herbert Allen,** Houston, Tex.

[73] Assignee: **Hallen Company,** Houston, Tex.

[21] Appl. No.: **925,365**

[22] Filed: **Jul. 17, 1978**

[51] **Int. Cl.**[3] **B67B 7/00**

[52] **U.S. Cl.** **81/3.36;** 81/3.48

[58] **Field of Search** .... 30/346, 346.53, 350; 81/3.1 A, 3.36, 3.37, 3.38 R, 3.38 A, 3.48

[56] **References Cited**

U.S. PATENT DOCUMENTS

279,203 6/1883 Strait .... 81/3.38 A

3,203,829 8/1965 Seyer et al. .... 30/346.53 X

FOREIGN PATENT DOCUMENTS

1048501 1/1959 Fed. Rep. of Germany .... 81/3.38 A

2340898 9/1977 France 81/3.48

Primary Examiner—James G. Smith

Attorney, Agent, or Firm—Browning, Bushman & Zamecki

[57] **ABSTRACT**

The invention pertains to an apparatus for extracting a cork from a bottle comprising a cork-engaging member, including a corkscrew, and a holder for aligning and guiding the corkscrew, via its own diameter, with respect to the bottle. The cork-engaging member is completely removable from the holder for independent use, and to permit the corkscrew point to be used to sever the metal foil, which covers the bottle top, and both the holder and cork engaging member to be more easily cleaned. The holder includes a plurality of longitudinally extending circumferentially spaced, radially deflectable gripping elements connected to and spaced downwardly from a guide which receives the corkscrew, and are all approximately equally resiliently biased to a normal unflexed radial position. One embodiment is capable of two alternative modes of operation. The corkscrew itself is improved by the provision of an outer layer of friction reducing material.

32 Claims, 12 Drawing Figures

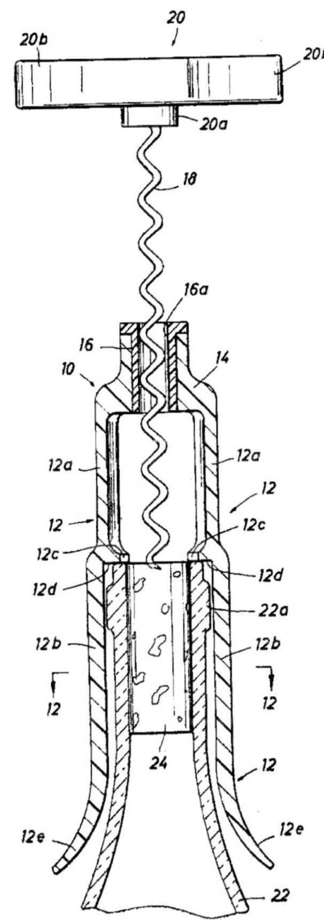

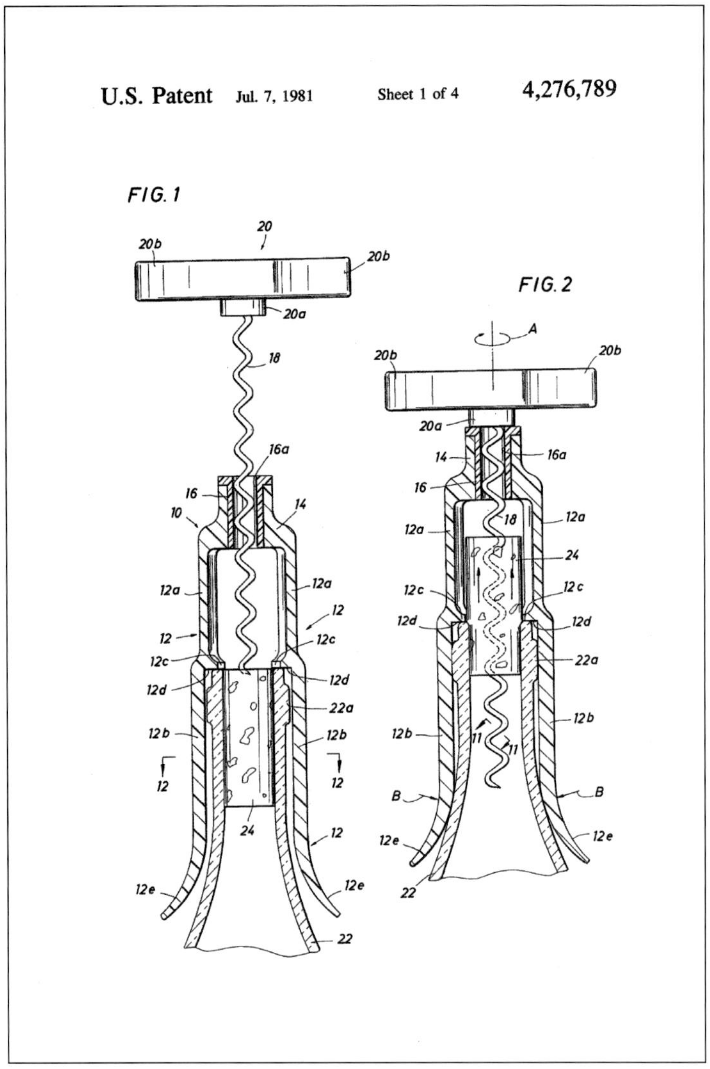
U.S. Patent Jul. 7, 1981 Sheet 1 of 4 4,276,789
FIG. 1
FIG. 2

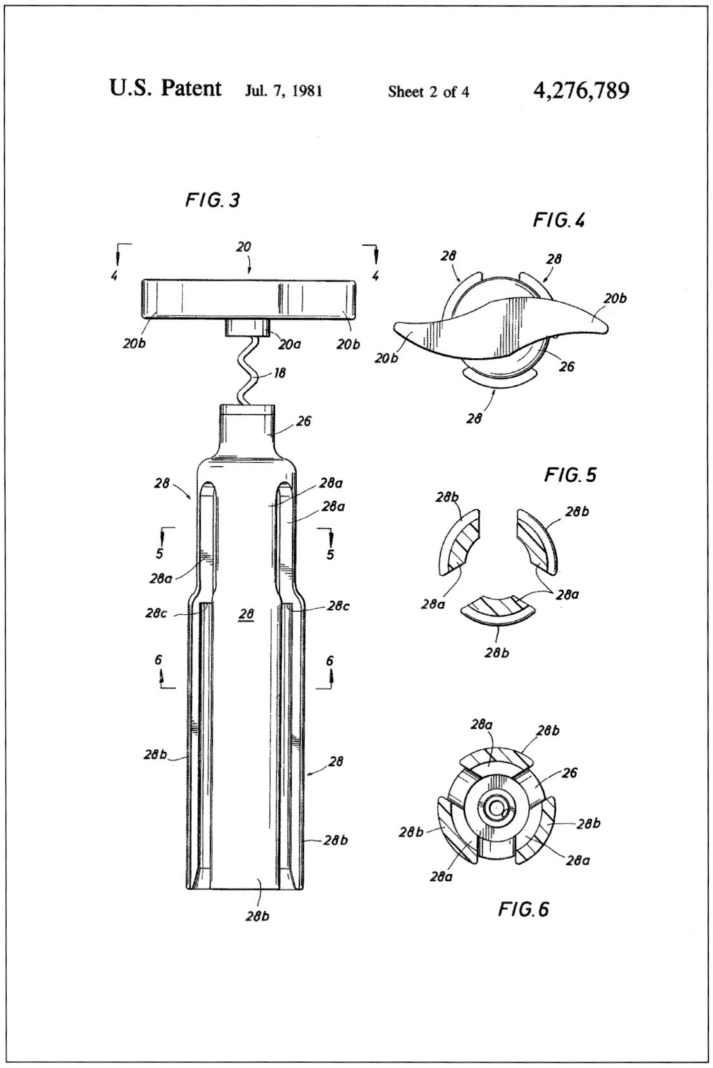
U.S. Patent Jul. 7, 1981 Sheet 2 of 4 4,276,789
FIG. 3
FIG. 4
FIG. 5
FIG. 6

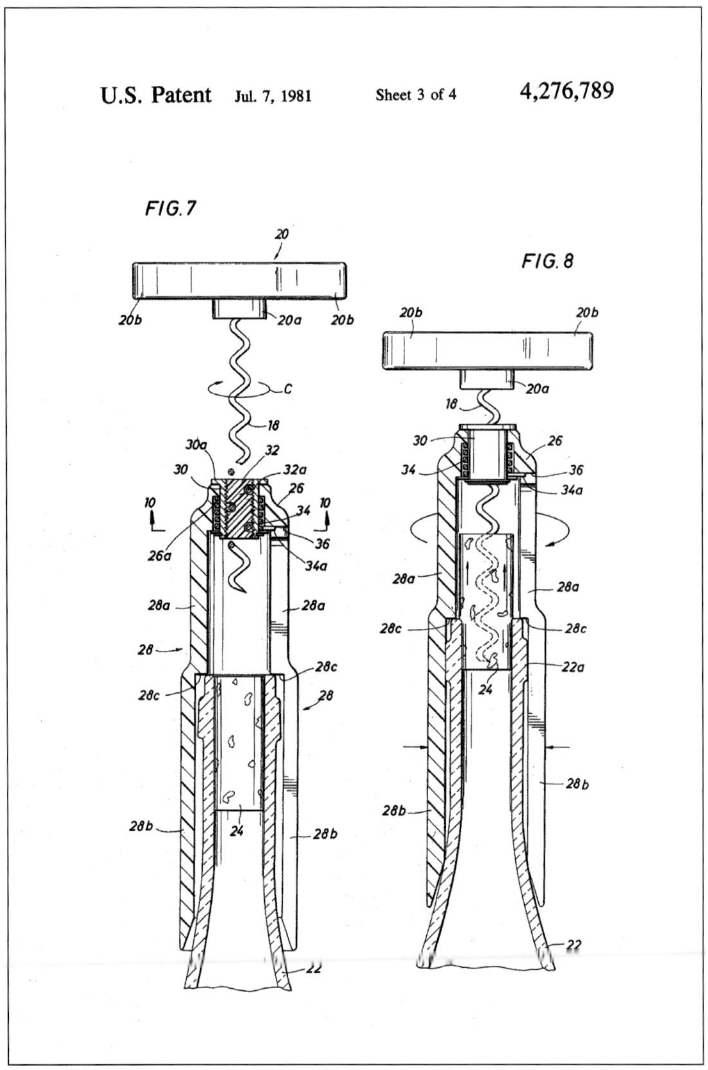
U.S. Patent Jul. 7, 1981 Sheet 3 of 4 4,276,789
FIG. 7
FIG. 8

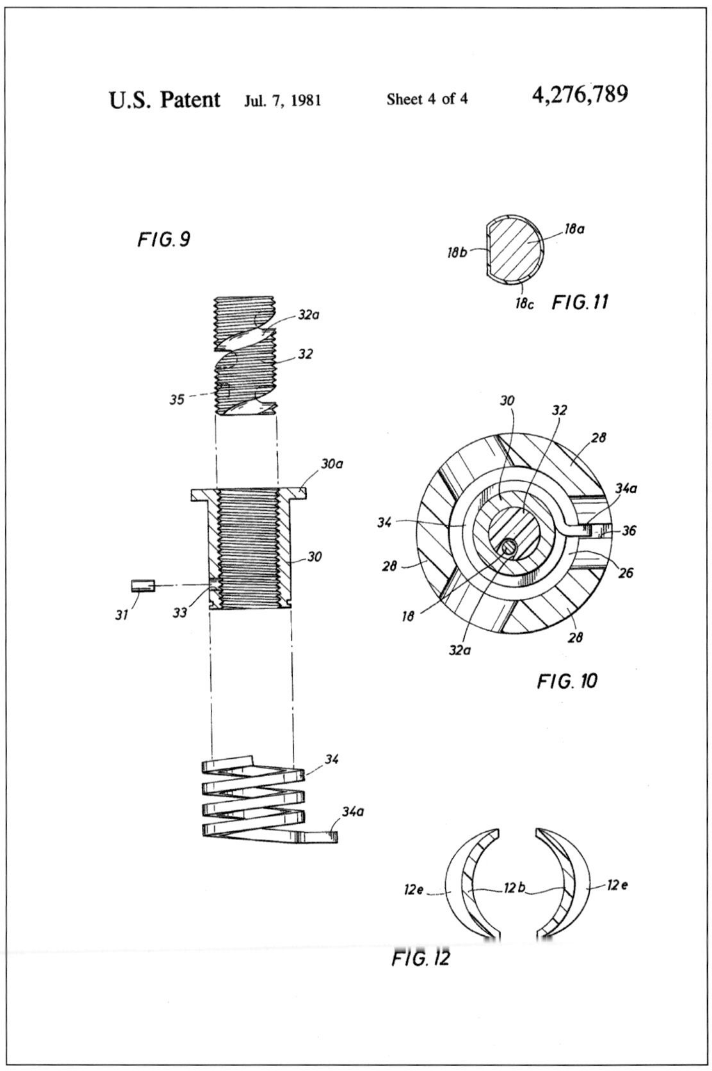
U.S. Patent Jul. 7, 1981 Sheet 4 of 4 4,276,789
FIG. 9
FIG. 11
FIG. 10
FIG. 12

United States Patent [19]

Allen

[11] **4,291,597**

[45] **Sep. 29, 1981**

[54] **CORK EXTRACTOR**

[75] Inventor: **Herbert Allen,** Houston, Tex.

[73] Assignee: **Hallen Company,** Houston, Tex.

[21] Appl. No.: **17,598**

[22] Filed: **Mar. 5, 1979**

Related U.S. Application Data

[63] Continuation-in-part of Ser. No. 925,365, Jul. 17, 1978.

[51] **Int. Cl.**³ **B67B 7/00**

[52] **U.S. Cl.** **81/3.48**

[58] **Field of Search** .... 30/346, 346.53, 350; 81/3.1 A, 3.36–3.38, 3.48

[56] **References Cited**

U.S. PATENT DOCUMENTS

279,203	6/1883	Strait	81/3.38 A
3,203,829	8/1965	Seyer et al.	30/346.53 X

FOREIGN PATENT DOCUMENTS

1048501	1/1954	Fed. Rep. of Germany	81/3.38 A
2340898	9/1977	France	81/3.48

Primary Examiner—James G. Smith

Attorney, Agent, or Firm—Browning, Bushman & Zamecki

[57] **ABSTRACT**

The invention pertains to an apparatus for extracting a cork from a bottle comprising a cork-engaging member, including a corkscrew, and a holder for aligning and guiding the corkscrew, via its own diameter, with respect to the bottle. The corkscrew itself is improved by the provision of an outer layer of friction reducing material such as a tetrafluoroethylene or other suitable plastic. Additionally, the tip portion of the corkscrew is formed in an improved manner which prevents fragments of cork from breaking off and falling into the bottle even though the screw is driven completely through the cork. Also disclosed is an improved method of making such a corkscrew.

7 Claims, 8 Drawing Figures

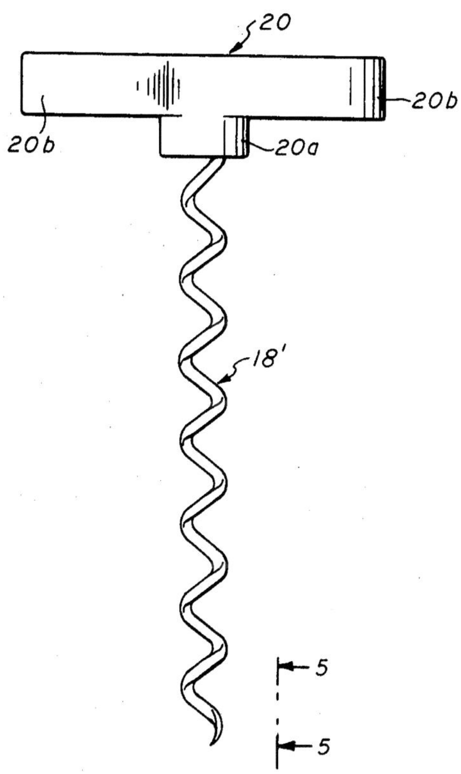

U.S. Patent Sep. 29, 1981 Sheet 1 of 2 4,291,597

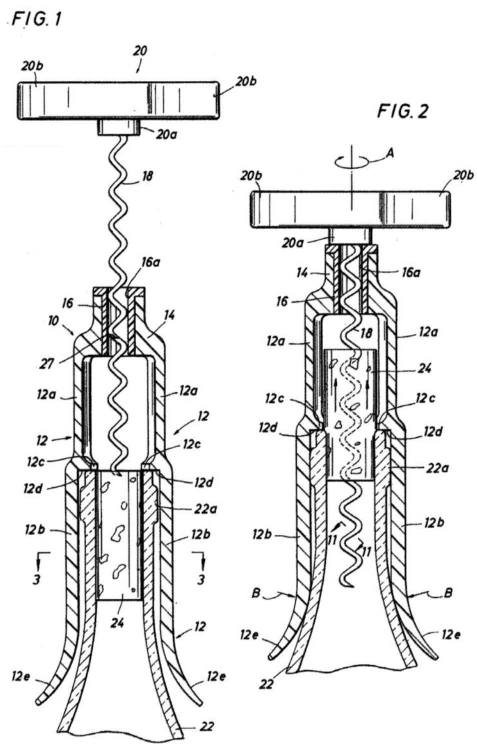

U.S. Patent Sep. 29, 1981 Sheet 2 of 2 4,291,597

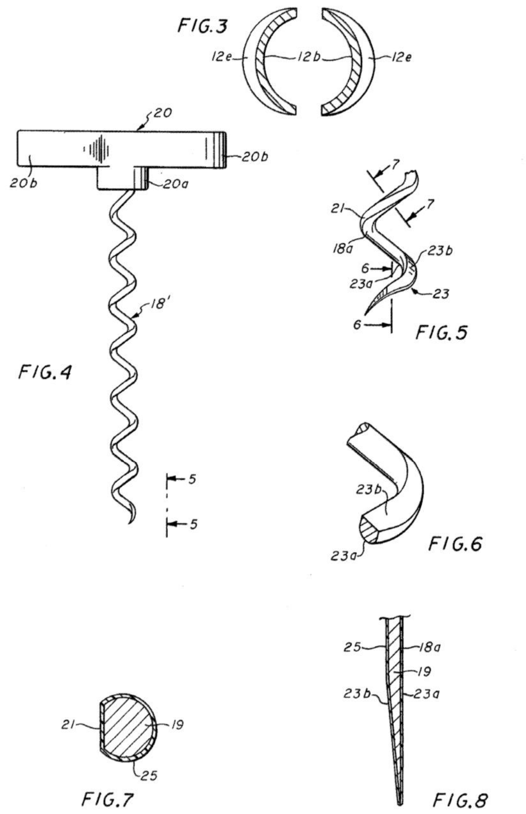

U.S. Patent Feb. 7, 1984 Sheet 1 of 2 4,429,444

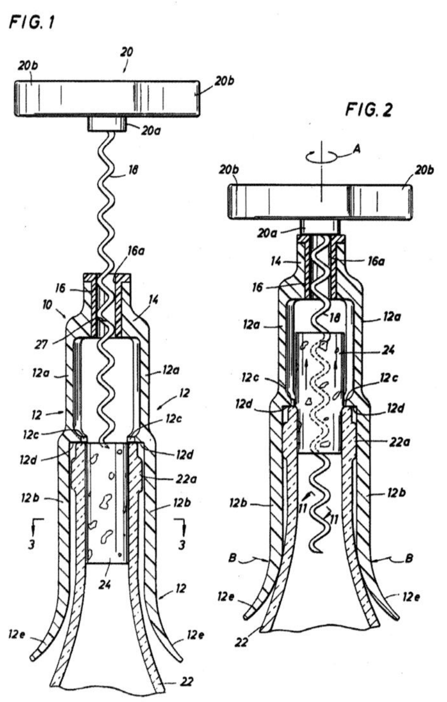

U.S. Patent Feb. 7, 1984 Sheet 2 of 2 4,429,444

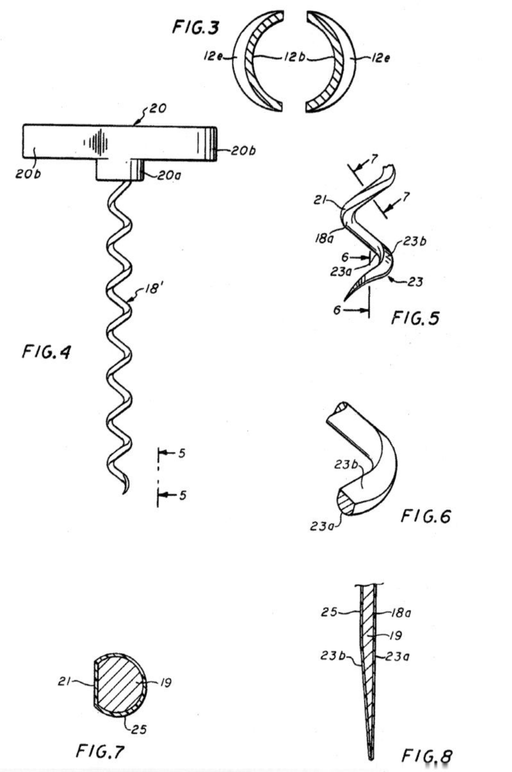

United States Patent [19]

Allen

[11] **4,429,444**

[45] **Feb. 7, 1984**

[54] **CORK EXTRACTOR**

[75] Inventor: **Herbert Allen,** Houston, Tex.

[73] Assignee: **Hallen Company,** Houston, Tex.

[21] Appl. No.: **245,584**

[22] Filed: **Mar. 19, 1981**

Related U.S. Application Data

[60] Division of Ser. No. 17,598, Mar. 5, 1979, Pat. No. 4,291,597, which is a continuation-in-part of Ser. No. 925,365, Jul. 17, 1978, Pat. No. 4,276,789.

[51] **Int. Cl.**³ **B23P 13/00**
[52] **U.S. Cl.** .... **29/33 F;** 140/86
[58] **Field of Search** .... 29/33 F, 557; 140/86; 81/3.45, 3.38, 3.48; 30/346.53

[56] **References Cited**

U.S. PATENT DOCUMENTS

441,137 11/1890 Clough 140/86
1,038,692 9/1912 Walker 140/86
3,203,829 8/1965 Seyer et al. .... 30/346.53 X

Primary Examiner—William R. Briggs
Attorney, Agent, or Firm—Browning, Bushman, Zamecki & Anderson

[57] **ABSTRACT**

The invention pertains to an apparatus for extracting a cork from a bottle comprising a cork-engaging member, including a corkscrew, and a holder for aligning and guiding the corkscrew, via its own diameter, with respect to the bottle. The corkscrew itself is improved by the provision of an outer layer of friction reducing material such as a tetrafluoroethylene or other suitable plastic. Additionally, the tip portion of the corkscrew is formed in an improved manner which prevents fragments of cork from breaking off and falling into the bottle even though the screw is driven completely through the cork. Also disclosed is an improved method of making such a corkscrew.

5 Claims, 8 Drawing Figures

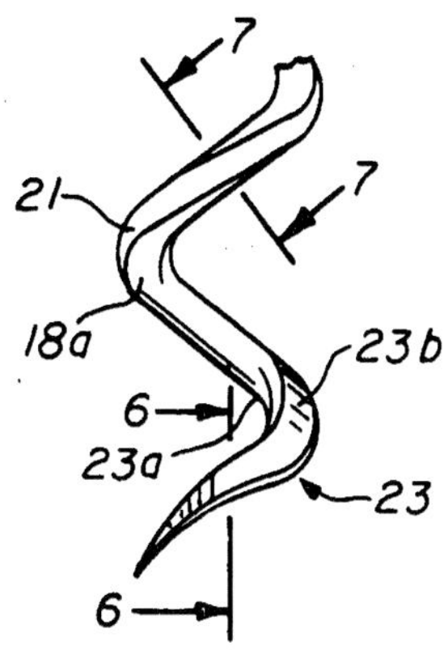

United States Patent [19]

Allen

[11] **4,377,096**

[45] *** Mar. 22, 1983**

[54] **CORK EXTRACTOR**

[75] Inventor: **Herbert Allen,** Houston, Tex.

[73] Assignee: **Hallen Company,** Houston, Tex.

[*] Notice: The portion of the term of this patent subsequent to Jul. 7, 1998, has been disclaimed.

[21] Appl. No.: **223,170**

[22] Filed: **Jan. 7, 1981**

Related U.S. Application Data

[63] Continuation-in-part of Ser. No. 17,598, Mar. 5, 1979, Pat. No. 4,291,597, which is a continuation-in-part of Ser. No. 925,365, Jul. 17, 1978, Pat. No. 4,276,789.

[51] **Int. Cl.**[3] **B67B 7/00**
[52] **U.S. Cl.** .... **81/3.38 A**
[58] **Field of Search** .... 81/3.36–3.38 A, 81/3.48

[56] **References Cited**

U.S. PATENT DOCUMENTS

32,396	5/1861	Burgess	81/3.38 A
206,134	7/1878	Richardson et al.	81/3.37 X
2,496,672	2/1950	Newman	81/3.38 R X
4,276,789	7/1981	Allen	81/3.36

FOREIGN PATENT DOCUMENTS

120684	6/1901	Switzerland	81/3.37

Primary Examiner—James G. Smith
Attorney, Agent, or Firm—Browning, Bushman, Zamecki & Anderson

[57] **ABSTRACT**

The invention comprises a cork extractor including a corkscrew and a guide receiving the corkscrew and permitting longitudinal and rotative movement of the corkscrew. A bottle-engaging device is connected to the guide for positioning the guide and corkscrew in generally coaxial alignment with the neck of the bottle, and this device includes a stop for limiting downward movement of the guide with respect to the bottle. Spacers interconnect the guide and the bottle-engaging device to space the guide upwardly from the bottle-engaging device. The spacers also define an opening for receipt of the cork as it emerges from the bottle. The invention further includes catches adjacent the spacers in the opening engageable with the cork as it emerges from the bottle to prevent rotation of the cork. An abutment carried by the corkscrew limits downward movement of the corkscrew with respect to the guide.

25 Claims, 16 Drawing Figures

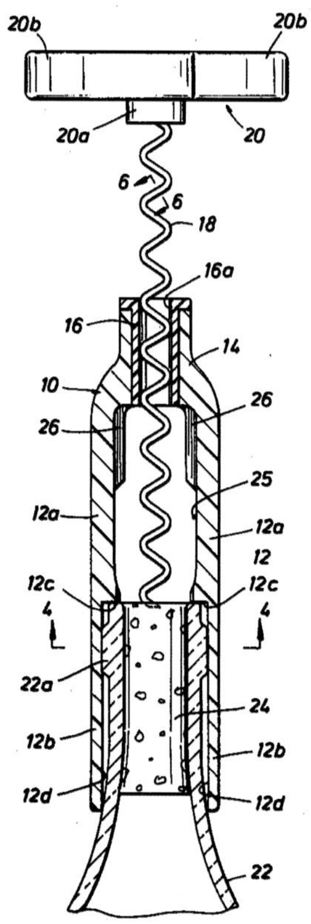

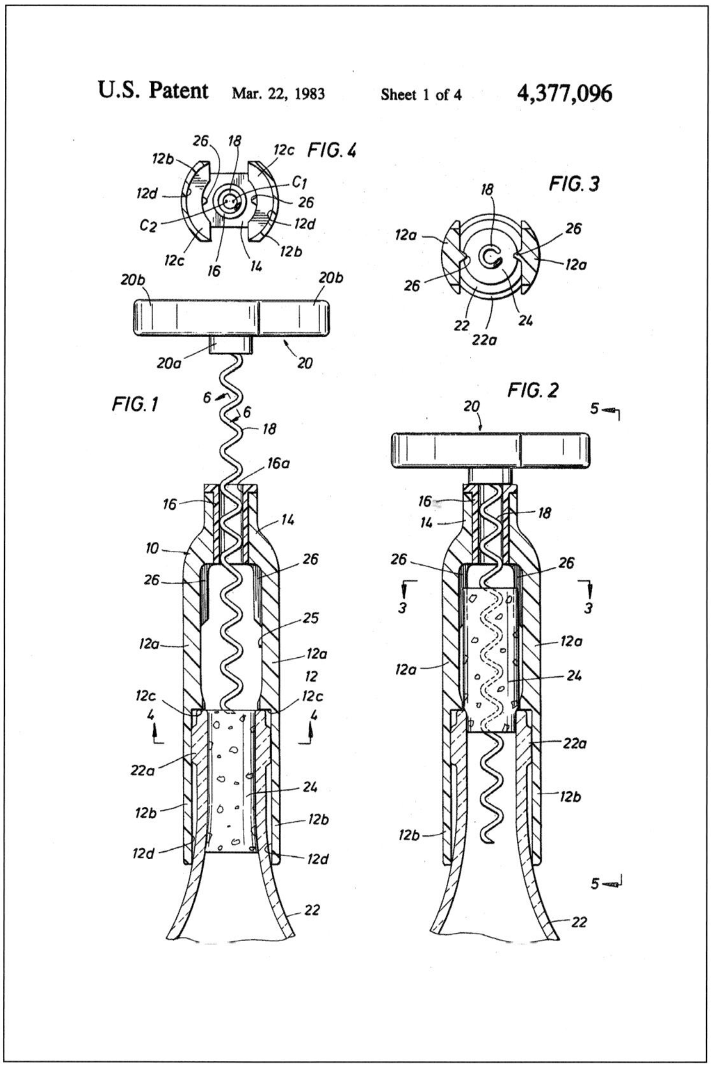
U.S. Patent Mar. 22, 1983 Sheet 1 of 4 4,377,096
FIG. 4
FIG. 3
FIG. 1
FIG. 2

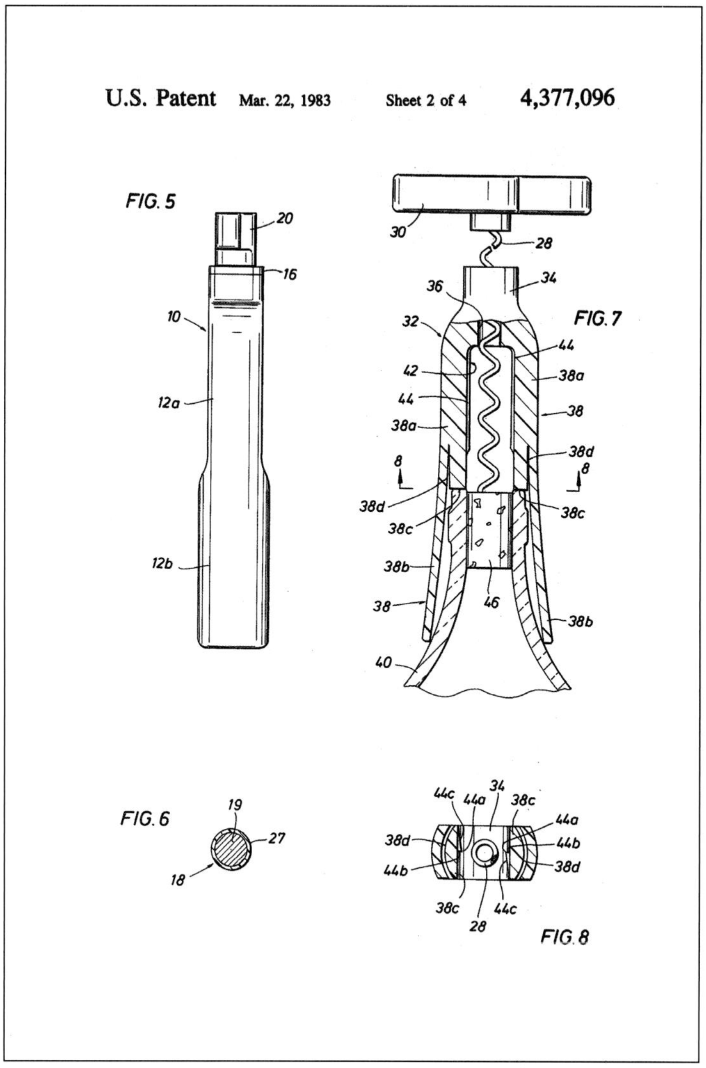
U.S. Patent Mar. 22, 1983 Sheet 2 of 4 4,377,096
FIG. 5
FIG. 7
FIG. 6
FIG. 8

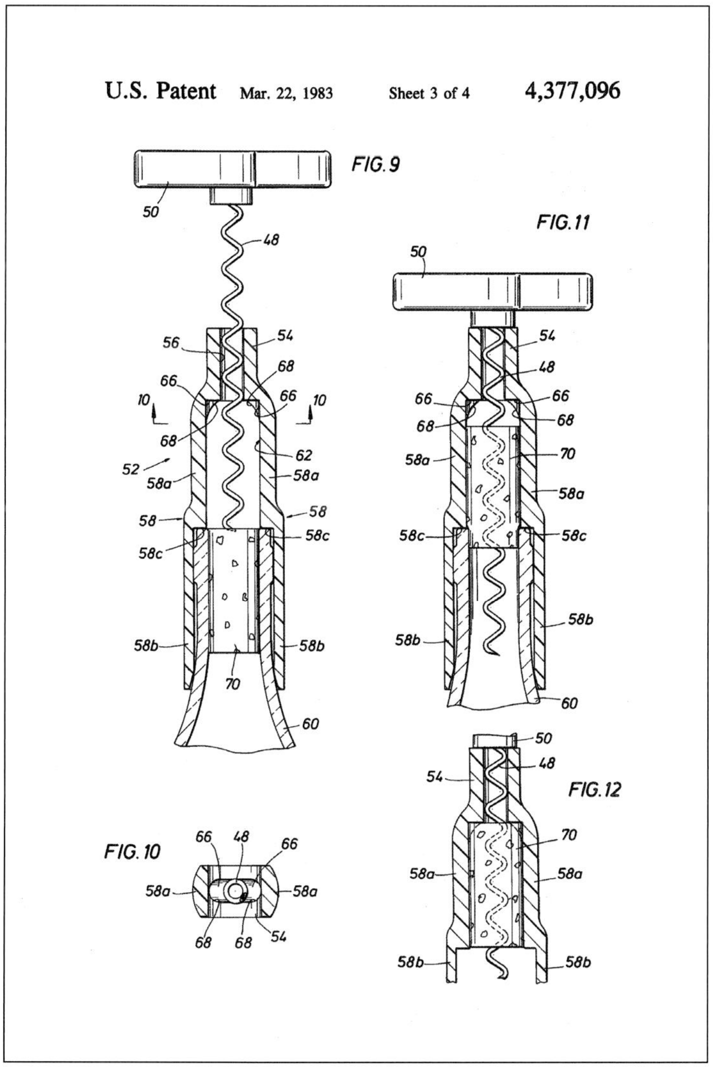
U.S. Patent Mar. 22, 1983 Sheet 3 of 4 4,377,096
FIG. 9
FIG. 11
FIG. 10
FIG. 12

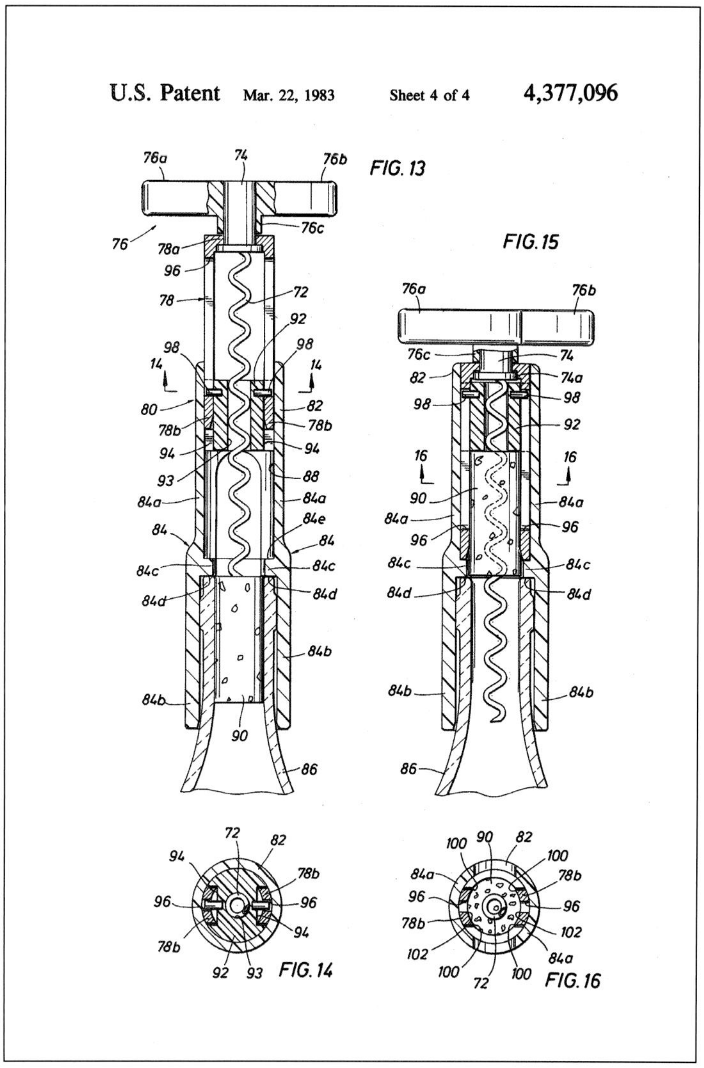
U.S. Patent Mar. 22, 1983 Sheet 4 of 4 4,377,096
FIG. 13
FIG. 15
FIG. 14
FIG. 16

United States Patent [19]

Allen

[11] **Patent Number: 4,703,673**

[45] **Date of Patent: Nov. 3, 1987**

[54] **CORK-EXTRACTING APPARATUS**

[75] Inventor: **Herbert Allen,** Houston, Tex.

[73] Assignee: **Hallen Company,** Houston, Tex.

[21] Appl. No.: **721,235**

[22] Filed: **Apr. 8, 1985**

[51] **Int. Cl.**[4] **B67B 7/04**

[52] **U.S. Cl.** **81/3.29;** 81/3.45

[58] **Field of Search** 81/3.29, 3.34, 3.48, 81/177.2, 3.45; 16/115

[56] **References Cited**

U.S. PATENT DOCUMENTS

35,362	5/1862	Chinnock .	
54,039	4/1866	Swartz .	
73,677	1/1868	Twigg .	
172,868	2/1876	Clough .	
181,147	8/1876	Decker .	
276,804	5/1883	Green .	
427,099	5/1890	Middlekauff .	
518,018	4/1894	Becker .	
814,834	3/1906	Coughlin .	
2,382,291	8/1945	Carlberg	16/115 X
2,980,456	4/1961	McMullin	16/115 X
3,256,756	6/1966	Del Piccolo	87/3.48
4,276,789	7/1981	Allen .	
4,291,597	9/1981	Allen .	
4,377,096	3/1983	Allen .	
4,429,444	2/1984	Allen .	

FOREIGN PATENT DOCUMENTS

705901	6/1931	France	81/3.29
1201854	7/1959	France .	
246954	11/1947	Switzerland	81/3.48
257086	3/1949	Switzerland	81/3.48

OTHER PUBLICATIONS

Corkscrews for Collectors, Bernard M. Watney and Homer D. Babbidge Sotheby Parke Bernet, New York, 1981, pp. 17, 20–28, 37, 40–41, 61, 65, 98, 101 and 105.

The Wonderful Screwpull Brochure, Hallen Company, Sep., 1982.

Primary Examiner—Roscoe V. Parker
Attorney, Agent, or Firm—Browning, Bushman, Zamecki & Anderson

[57] **ABSTRACT**

Apparatus for extracting a cork from a bottle comprises a movable portion, a handle, and a holder. The movable portion includes a helical corkscrew. The handle includes at least one elongate arm and is associated with the movable portion such that the arm may extend radially outwardly with respect to the corkscrew in an operating position of the handle. The arm has a restraining formation disposed distal the corkscrew in the operating position and adapted to engage a human finger and restrain it against movement radially outwardly. The holder is engageable with a bottle and defines an opening for receipt of a cork as it emerges from the bottle. The holder has a guide which allows rotational and longitudinal movement of the movable portion relative to the holder whereby the corkscrew can be driven downwardly into the cork and to a lowered position with respect to the holder upon rotation of the movable portion in a first direction. The movable portion and the holder have force transmitting formations interengageable, when the movable portion is in its lowered position, and adapted to cause the cork to be removed from the bottle upon further rotation of the movable portion.

33 Claims, 18 Drawing Figures

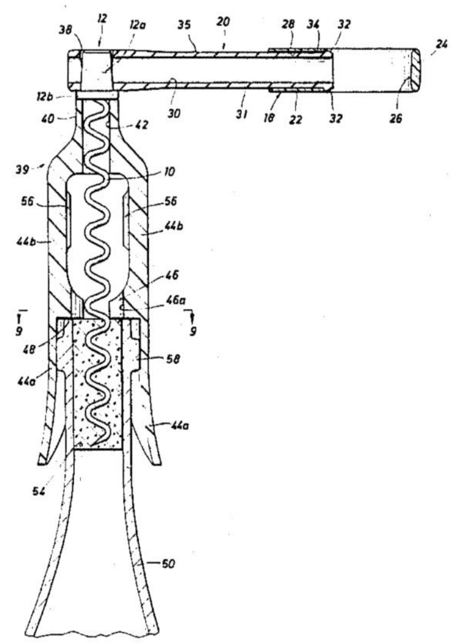

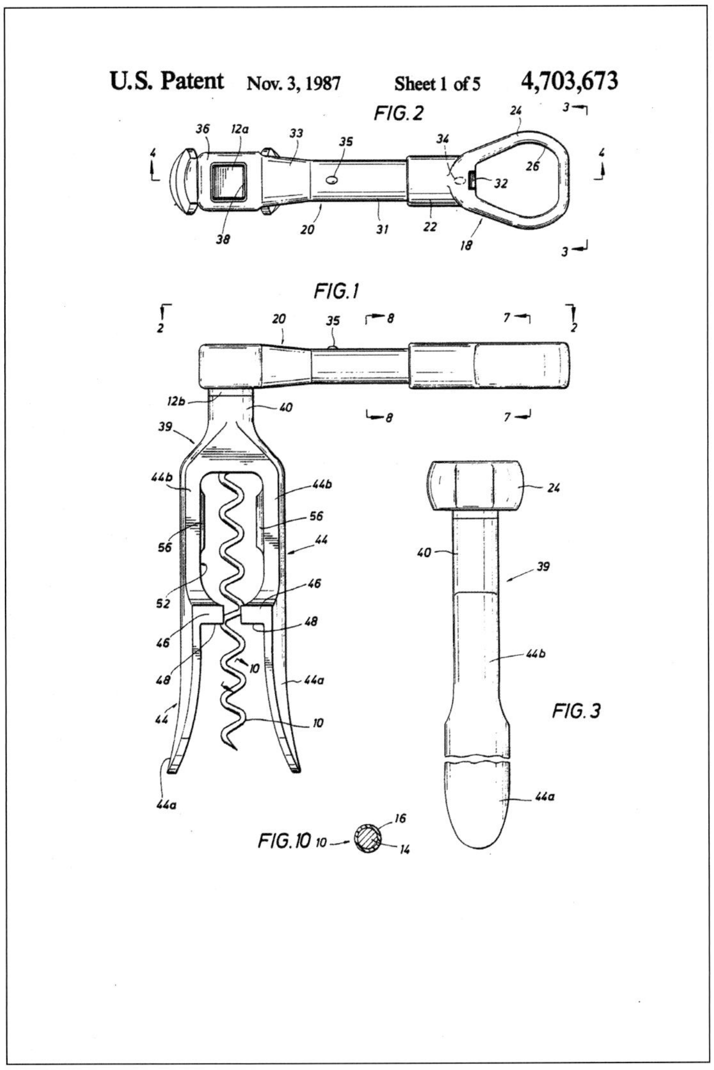
U.S. Patent Nov. 3, 1987 Sheet 1 of 5 4,703,673
FIG. 2
FIG. 1
FIG. 3
FIG. 10

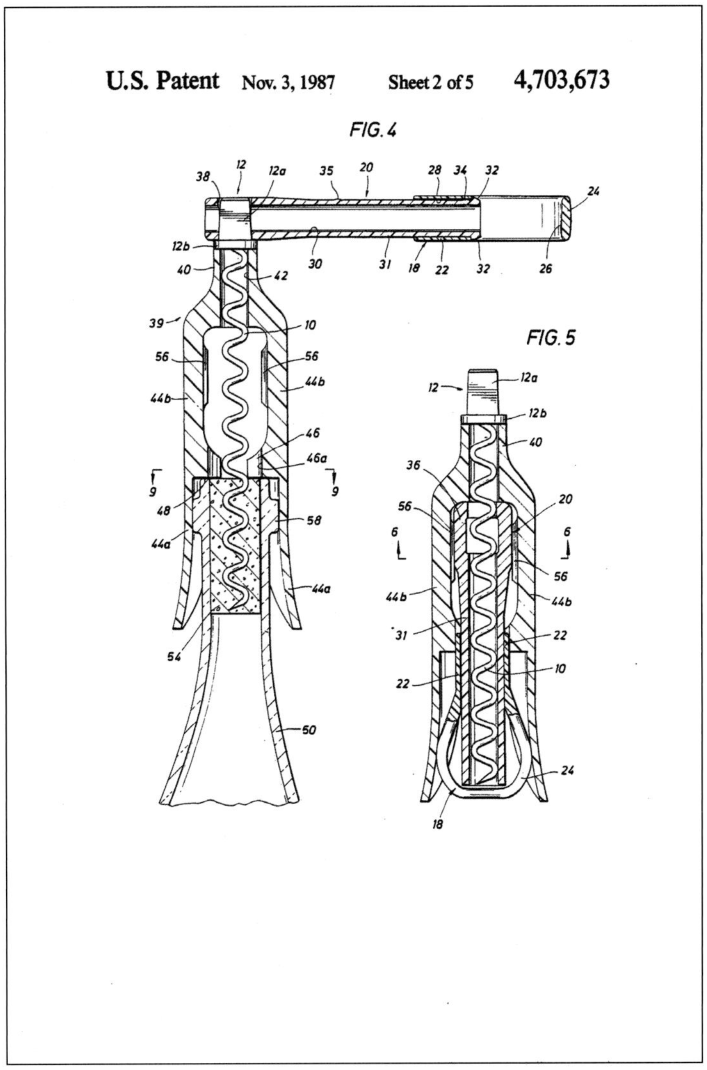
U.S. Patent Nov. 3, 1987 Sheet 2 of 5 4,703,673
FIG. 4
FIG. 5

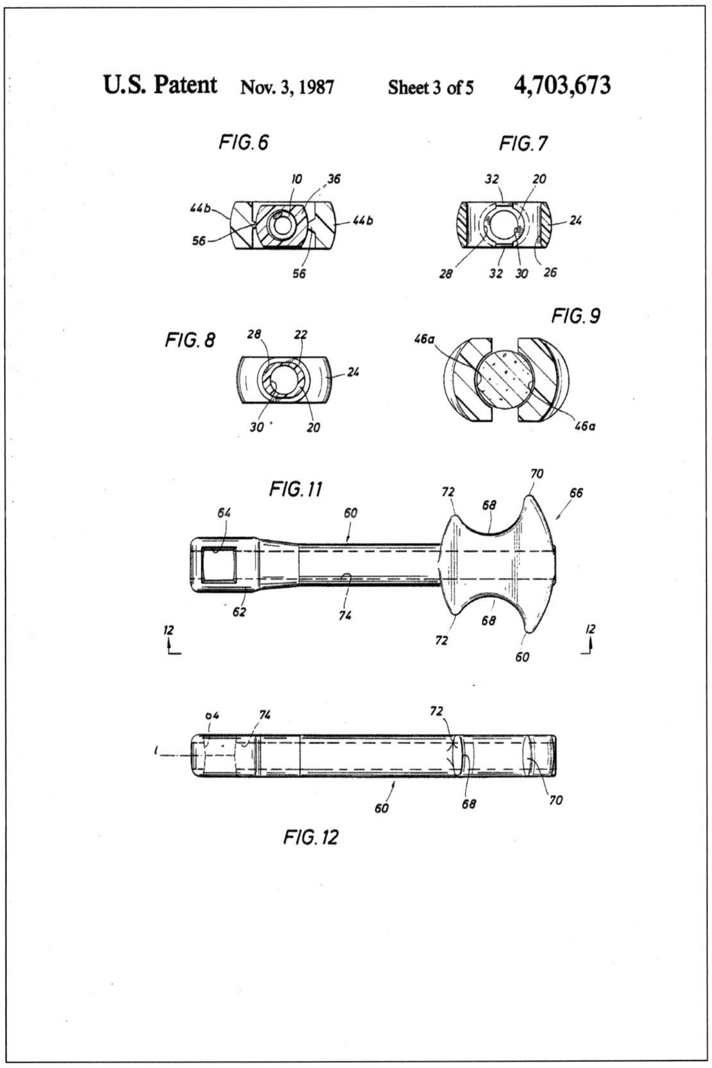
U.S. Patent Nov. 3, 1987 Sheet 3 of 5 4,703,673
FIG. 6
FIG. 7
FIG. 8
FIG. 9
FIG. 11
FIG. 12

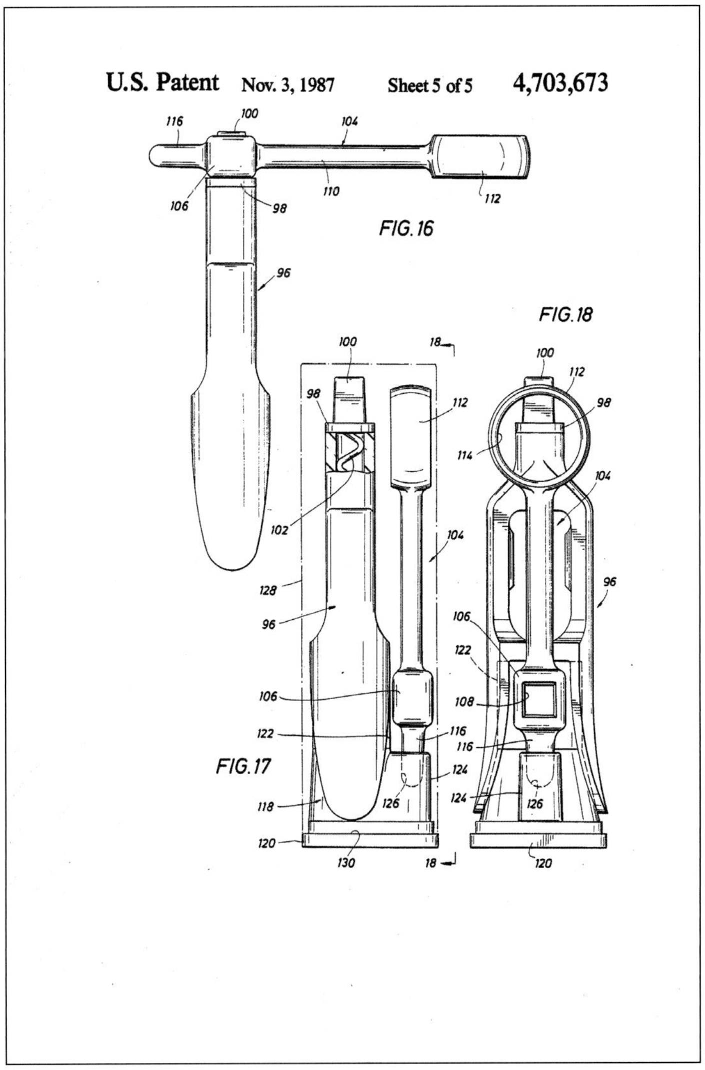
U.S. Patent Nov. 3, 1987 Sheet 5 of 5 4,703,673
FIG. 16
FIG. 18
FIG. 17

United States Patent [19]

Allen

[11] **Patent Number: 4,800,783**

[45] **Date of Patent: Jan. 31, 1989**

[54] **METHOD AND APPARATUS FOR REMOVING A CORK OR PLASTIC STOPPER FROM A CHAMPAGNE BOTTLE**

[76] Inventor: **Herbert Allen,** 3207 Groveland, Houston, Tex. 77019

[21] Appl. No.: **58,566**

[22] Filed: **Jun. 5, 1987**

[51] **Int. Cl.**[4] **B67B 7/02**

[52] **U.S. Cl.** **81/3.4**

[58] **Field of Search** 81/3.4, 3.07, 3.44

[56] **References Cited**

U.S. PATENT DOCUMENTS

2,495,308	1/1950	Amigone	81/3.44
2,761,338	12/1953	Hardy .	
3,628,405	12/1971	Fleisher .	
4,296,653	10/1981	Lundgren et al. .	
4,387,609	6/1983	Polsfuss .	
4,433,360	3/1984	Becker .	
4,442,735	4/1984	Chance et al. .	
4,474,087	10/1984	Widman .	
4,519,277	4/1985	Raab .	
4,602,245	8/1986	Veverka et al. .	

FOREIGN PATENT DOCUMENTS

1519035	2/1968	France .	
149173	3/1955	Sweden	81/3.4
10192	of 1889	United Kingdom	81/3.44
27632	of 1910	United Kingdom	81/3.44
83/02265	7/1983	World Int. Prop. O.	81/3.4

OTHER PUBLICATIONS

Advertisement for the Champagne Key, undated, one page.

Wine Ambiance, catalog by America's Wineland Crafts, 1986 Edition, p. 8.

The Wine Enthusiast, catalog, 1986 Edition, pp. 2 and 14.

Advertisement for the Uncorker, dated 10/83, one page.

Primary Examiner—Roscoe V. Parker

Attorney, Agent, or Firm—Browning, Bushman, Zamecki & Anderson

[57] **ABSTRACT**

Apparatus and method for removing a stopper from a bottle having an internal pressure greater than the ambient pressure surrounding the bottle, e.g. a champagne bottle. At least one generally vertical edge of the apparatus is placed in a generally vertical groove in the periphery of the stopper, in some cases cutting the groove simultaneously as it is moved downward over the stopper. The apparatus is then oscillated or rotated, whereupon the internal pressure in the bottle can overcome the sliding friction between the stopper and the bottle. The apparatus is also adapted to control the stopper, once expelled.

24 Claims, 3 Drawing Sheets

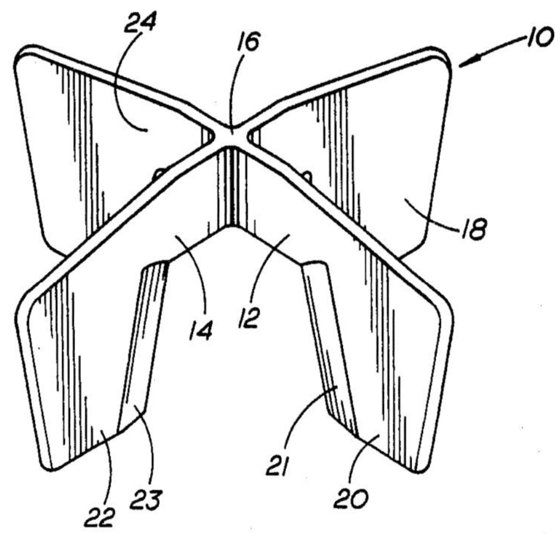

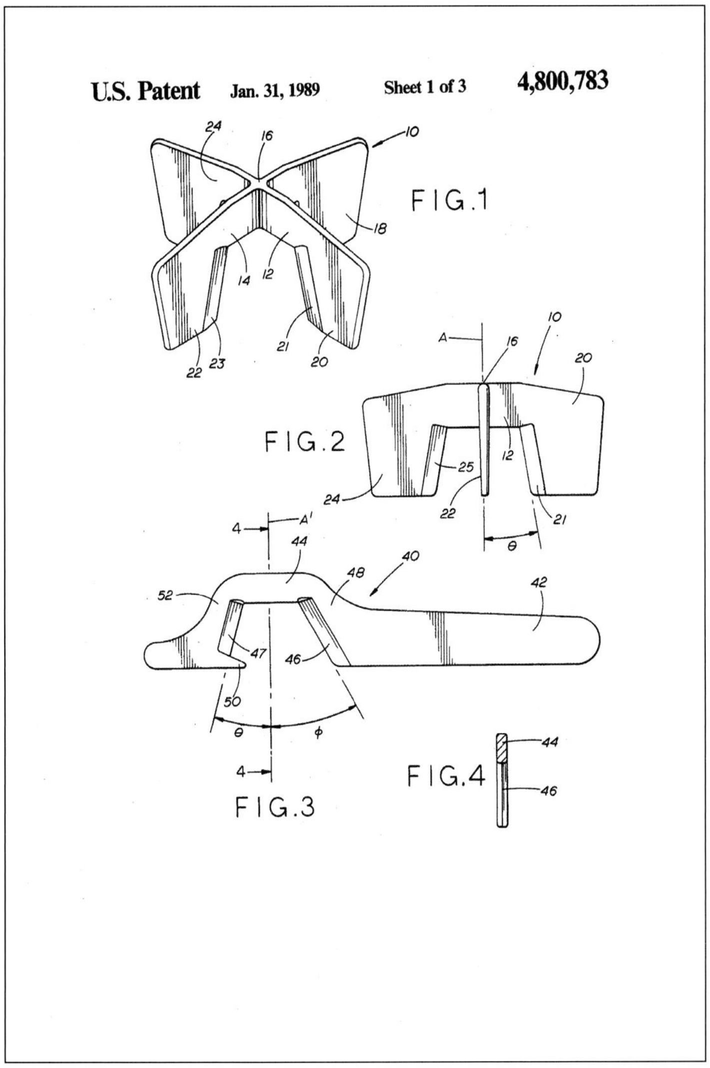
U.S. Patent Jan. 31, 1989 Sheet 1 of 3 4,800,783
FIG.1
FIG.2
FIG.3
FIG.4

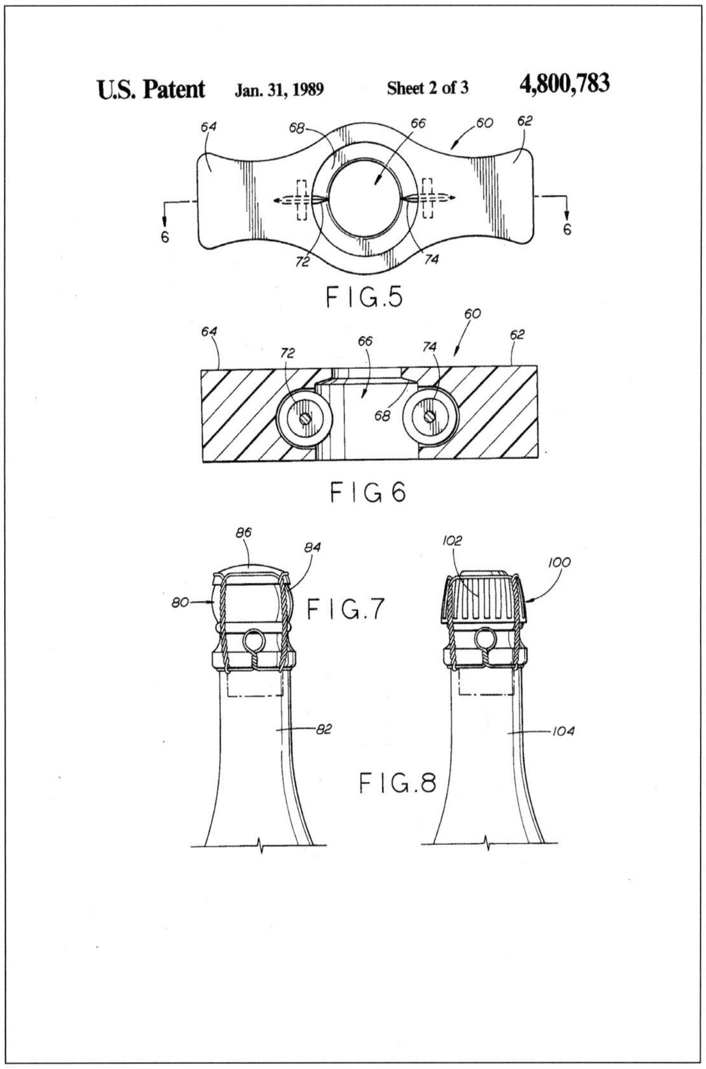
U.S. Patent Jan. 31, 1989 Sheet 2 of 3 4,800,783
FIG.5
FIG 6
FIG.7
FIG.8

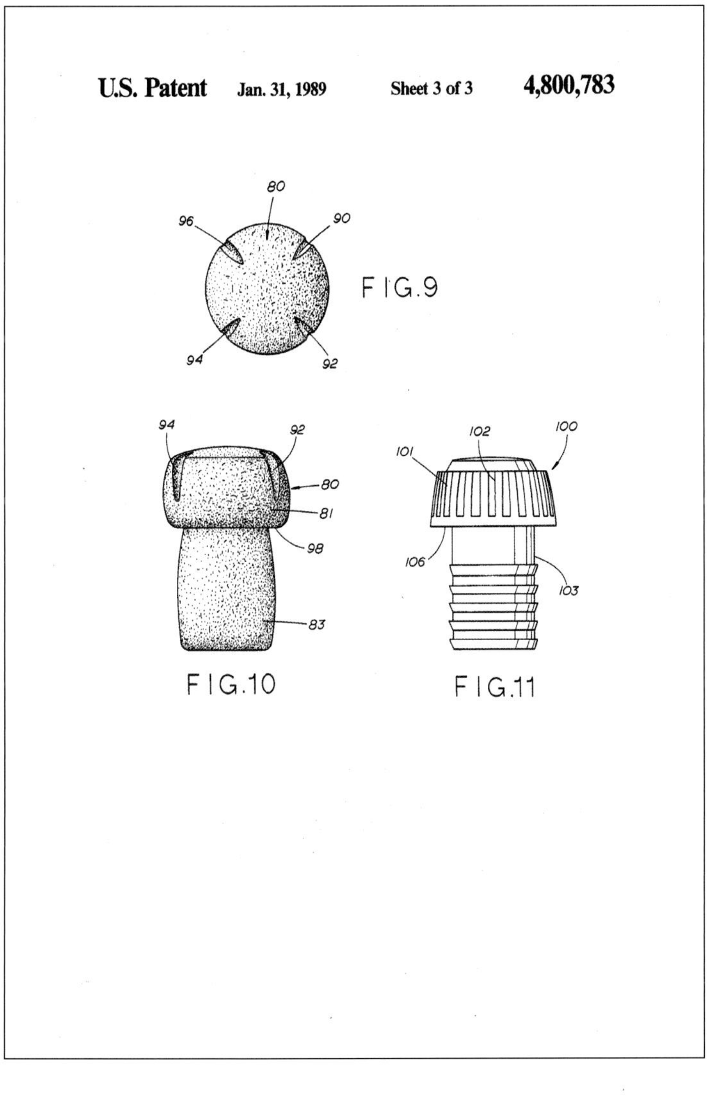
U.S. Patent Jan. 31, 1989 Sheet 3 of 3 4,800,783
FIG.9
FIG.10
FIG.11

United States Patent [19]

Allen

[11] **Patent Number: 4,800,784**

[45] **Date of Patent: Jan. 31, 1989**

[54] **APPARATUS FOR REMOVING CORKS FROM BOTTLES**

[75] Inventor: **Herbert Allen,** Houston, Tex.

[73] Assignee: **Hallen Company,** Houston, Tex.

[21] Appl. No.: **168,163**

[22] Filed: **Mar. 15, 1988**

[51] **Int. Cl.**[4] **B67B 7/02**
[52] **U.S. Cl.** **81/3.36;** 81/3.48
[58] **Field of Search** 81/3.36, 3.48, 3.07, 81/3.45

[56] **References Cited**

U.S. PATENT DOCUMENTS

4,276,789	7/1981	Allen	81/3.36
4,291,597	9/1981	Allen	81/3.48
4,377,096	3/1983	Allen	81/3.36
4,429,444	2/1984	Allen	140/86

Primary Examiner—Roscoe V. Parker
Attorney, Agent, or Firm—Browning, Bushman, Zamecki & Anderson

[57] **ABSTRACT**

An apparatus for removing corks from bottles is adapted to allow for automatic ejection of the cork from the apparatus, after the cork has been withdrawn from the bottle, by reverse rotation of the corkscrew. The apparatus comprises a corkscrew having a first helical section with a first outer diameter and a second helical section disposed below the first section and with a second outer diameter greater than the first outer diameter so that an upwardly directed screw shoulder is formed between the sections. A handle is fixed to the upper end of the corkscrew. The corkscrew is longitudinally movable with respect to a guide frame therefor. The guideframe includes stops for abutting the top of a bottle and spacers extending upwardly from the stops and defining a cork receiving space for receipt of a cork as it is withdrawn from a bottle by the apparatus. Catches are connected to the spacers, and disposed in the cork receiving space, the catches being engageable with a cork to prevent rotation thereof. Upward movement of the corkscrew shoulder with respect to the spacer means is limited by the screw.

32 Claims, 4 Drawing Sheets

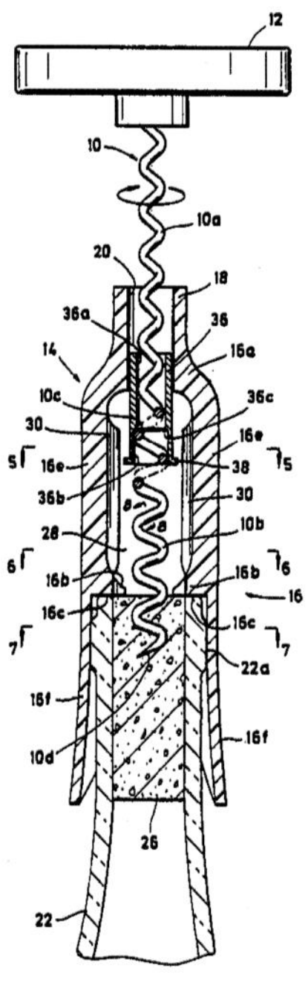

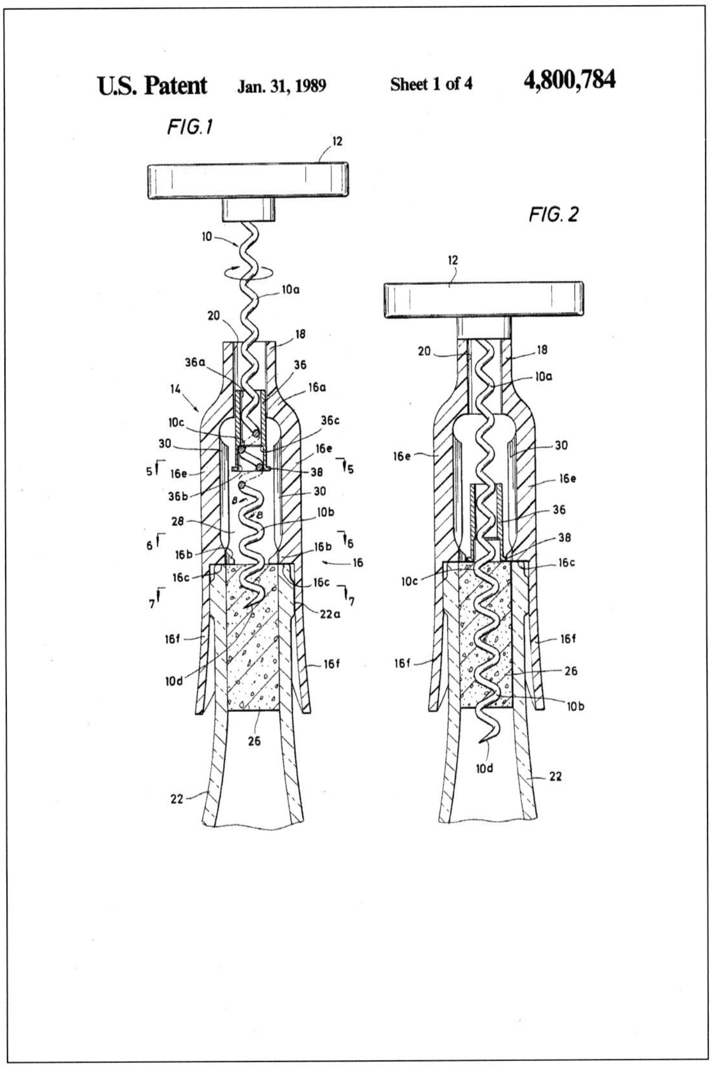
U.S. Patent Jan. 31, 1989 Sheet 1 of 4 4,800,784
FIG. 1
FIG. 2

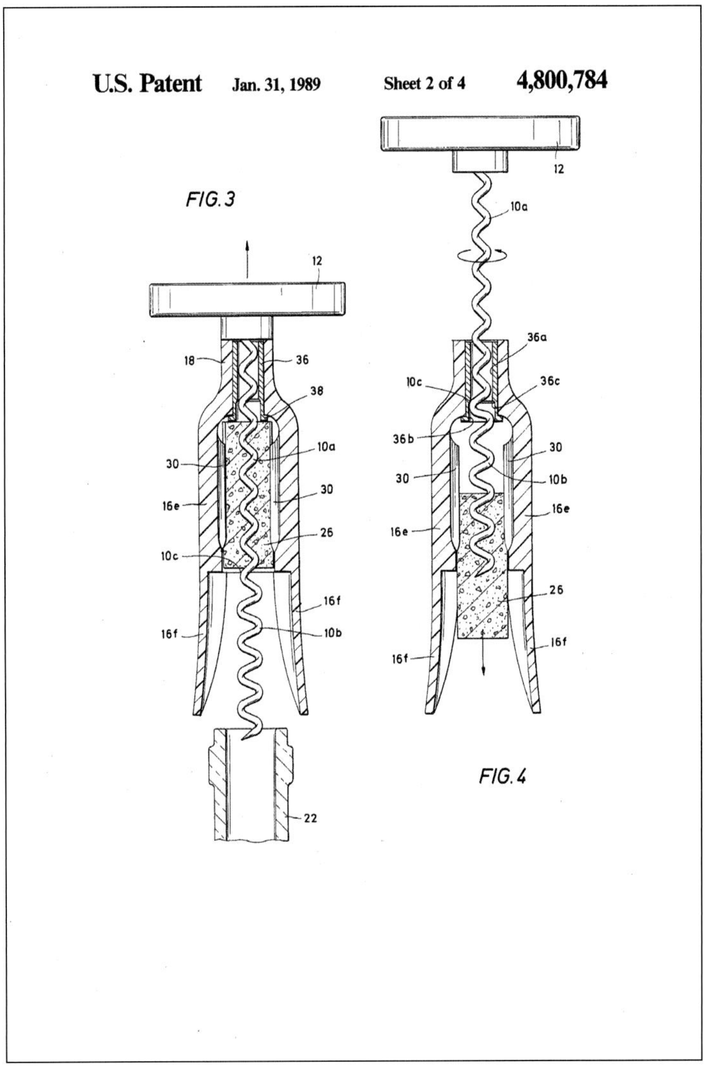
U.S. Patent Jan. 31, 1989 Sheet 2 of 4 4,800,784
FIG. 3
FIG. 4

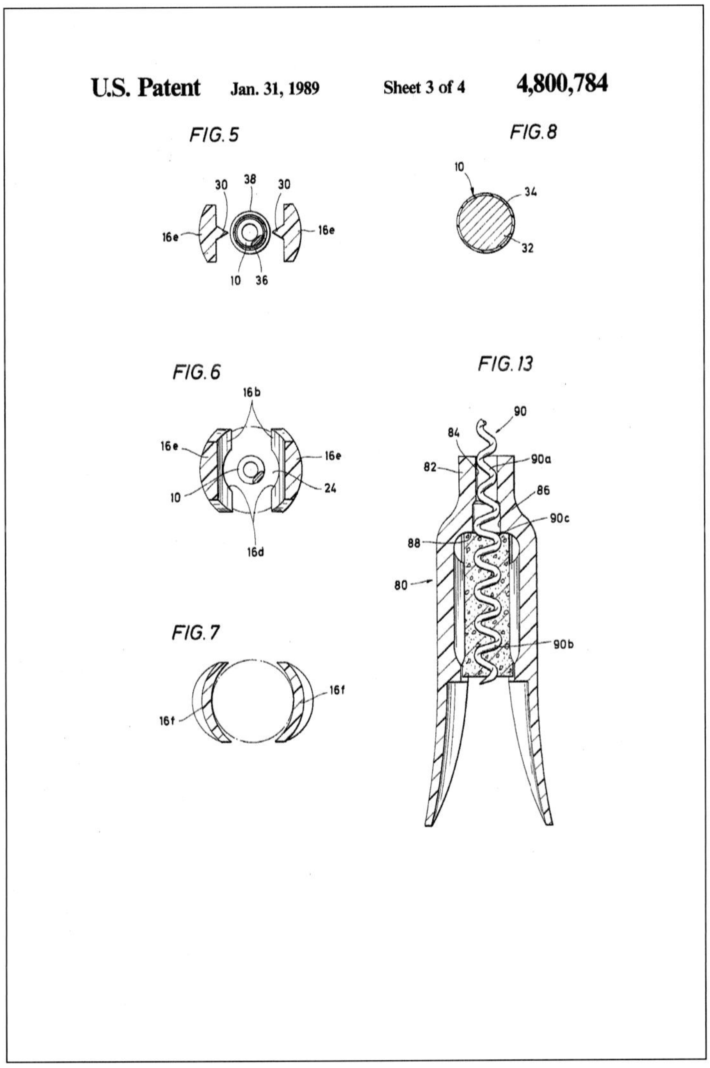
U.S. Patent Jan. 31, 1989 Sheet 3 of 4 4,800,784
FIG. 5
FIG. 8
FIG. 6
FIG. 13
FIG. 7

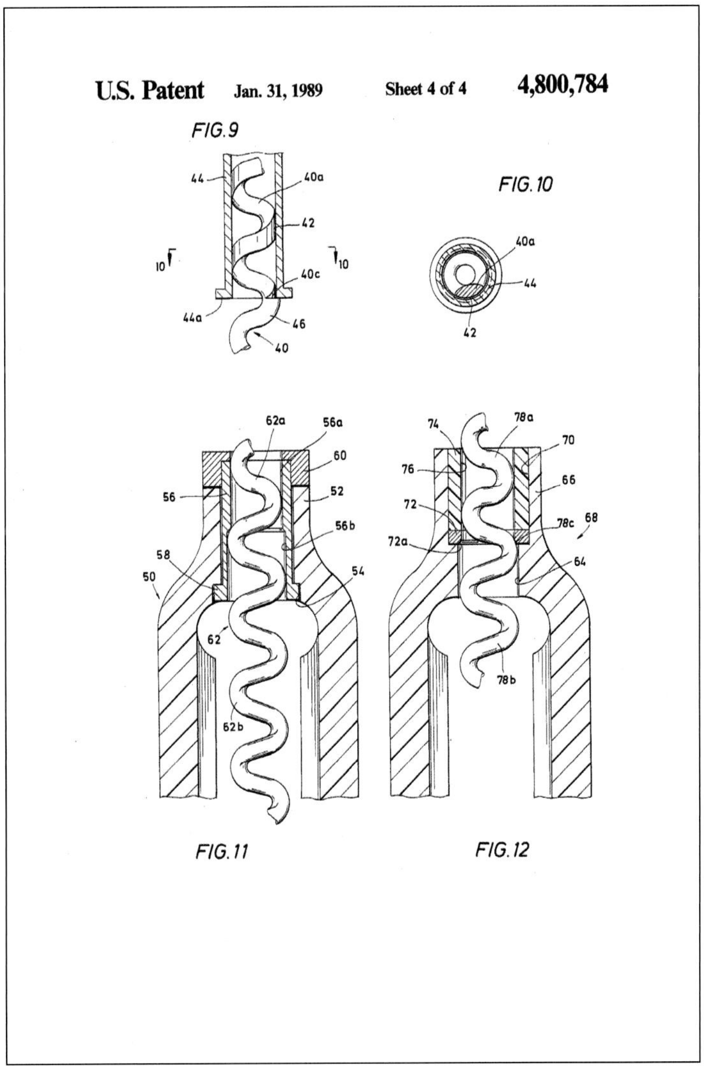
U.S. Patent Jan. 31, 1989 Sheet 4 of 4 4,800,784
FIG. 9
FIG. 10
FIG. 11
FIG. 12

United States Patent [19]

Allen

[11] **Patent Number: 4,845,844**

[45] **Date of Patent: Jul. 11, 1989**

[54] **FOIL CUTTER**

[75] Inventor: **Herbert Allen,** Houston, Tex.

[73] Assignee: **Hallen Company,** Houston, Tex.

[21] Appl. No.: **238,988**

[22] Filed: **Aug. 24, 1988**

Related U.S. Application Data

[63] Continuation of Ser. No. 790,263, Oct. 22, 1985, abandoned.

[51] **Int. Cl.**[4] **B67B 7/00**
[52] **U.S. Cl.** **30/1.5;** 30/102
[58] **Field of Search** 30/1.5, 101, 102, 253; 7/156

[56] **References Cited**

U.S. PATENT DOCUMENTS

D. 142,202	8/1945	Lanyon	30/1.5
896,303	8/1908	Klepper	30/1.5
996,433	6/1911	Stone	30/1.5
999,668	8/1911	Montaperto	30/102
1,102,124	6/1914	Baptiste	30/1.5
1,531,903	3/1925	Cummins	30/253 X
2,439,894	4/1948	Jahn et al.	30/1.5
2,503,517	4/1950	Sirica .	
2,566,653	9/1951	Callahan	30/1.5
2,589,742	3/1952	Smith	30/2
2,681,504	6/1954	Fox .	
2,694,854	11/1954	Kannenberg	30/1.5
2,718,692	9/1955	Krzanowski	30/1.5
2,823,395	2/1958	Brownson	30/1.5 X
3,025,597	3/1962	Huglin	30/102 X

FOREIGN PATENT DOCUMENTS

839319	3/1952	Fed. Rep. of Germany	30/1.5
1097917	2/1955	France	30/1.5
15940	of 1889	United Kingdom	30/1.5

Primary Examiner—Douglas D. Watts
Attorney, Agent, or Firm—Browning, Bushman, Zambecki & Anderson

[57] **ABSTRACT**

A cutting device for cutting material covering wine bottle necks comprising cutters defining at least three cutting surfaces and a frame carrying the cutters. The frame has an axis and positions the cutters such that the cutting surfaces are radially spaced from the axis, aligned transversely in the axis and circumferentially spread with the angular displacement between each cutting surface and the two adjacent cutting surfaces less than 180°, whereby, upon rotation of the frame about the axis by an amount less than 180°, the cutting surfaces respectively define contiguous arcs of a common circle. The frame further permits the cutting surfaces to so define a plurality of such common circles of different diameters.

31 Claims, 5 Drawing Sheets

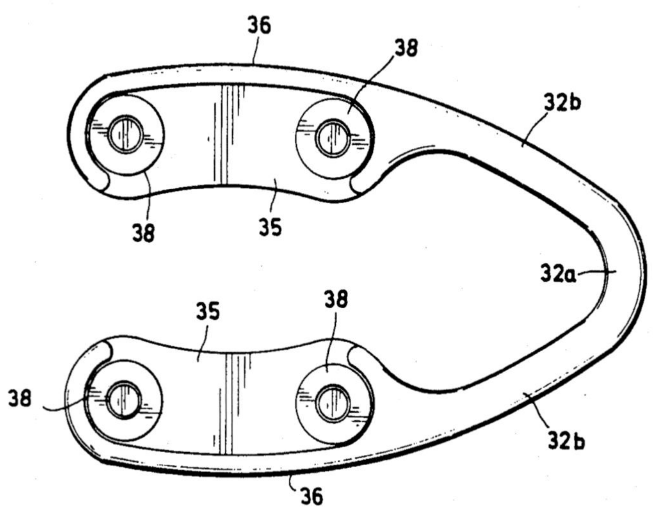

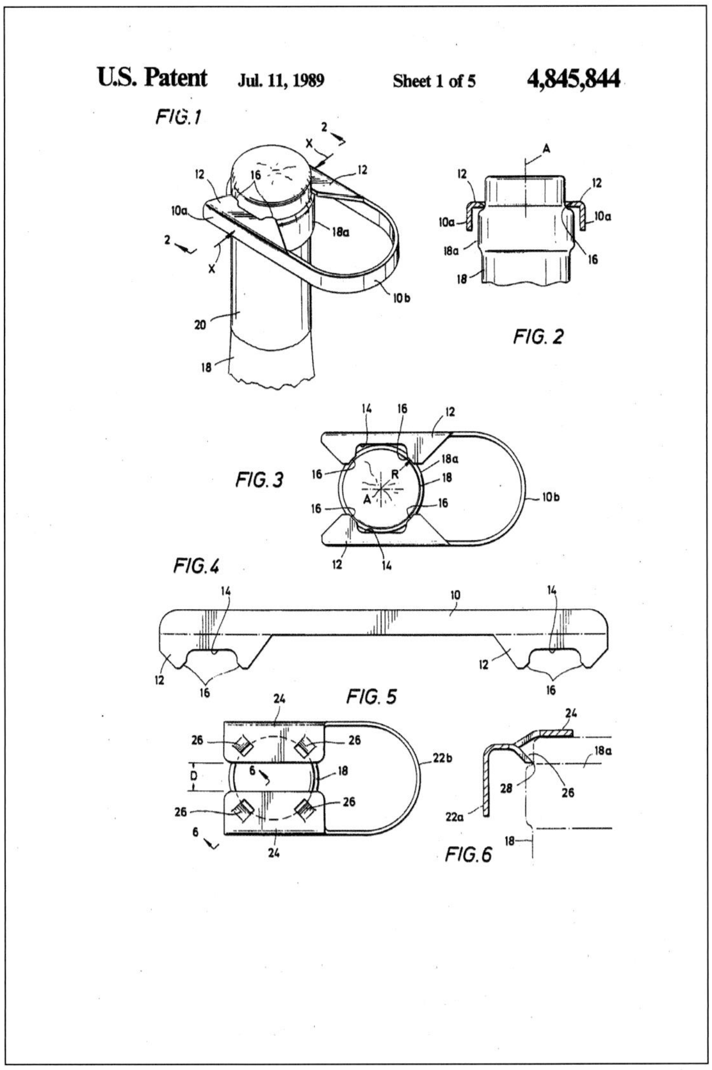
U.S. Patent Jul. 11, 1989 Sheet 1 of 5 4,845,844
FIG. 1
FIG. 2
FIG. 3
FIG. 4
FIG. 5
FIG. 6

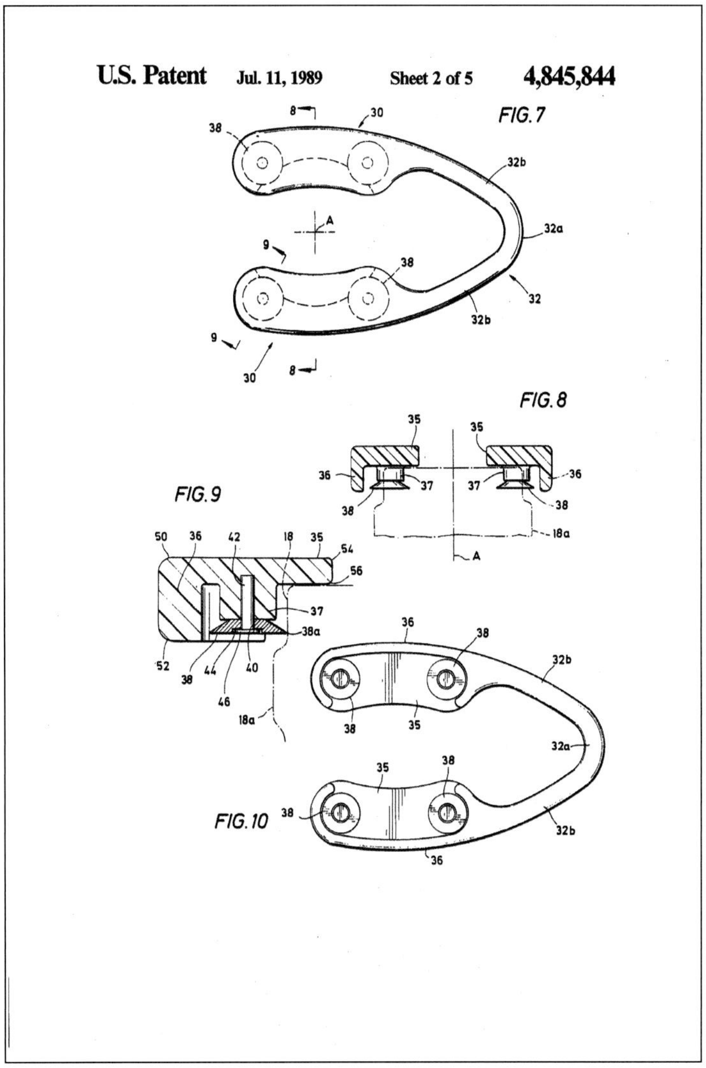
U.S. Patent Jul. 11, 1989 Sheet 2 of 5 4,845,844
FIG. 7
FIG. 8
FIG. 9
FIG. 10

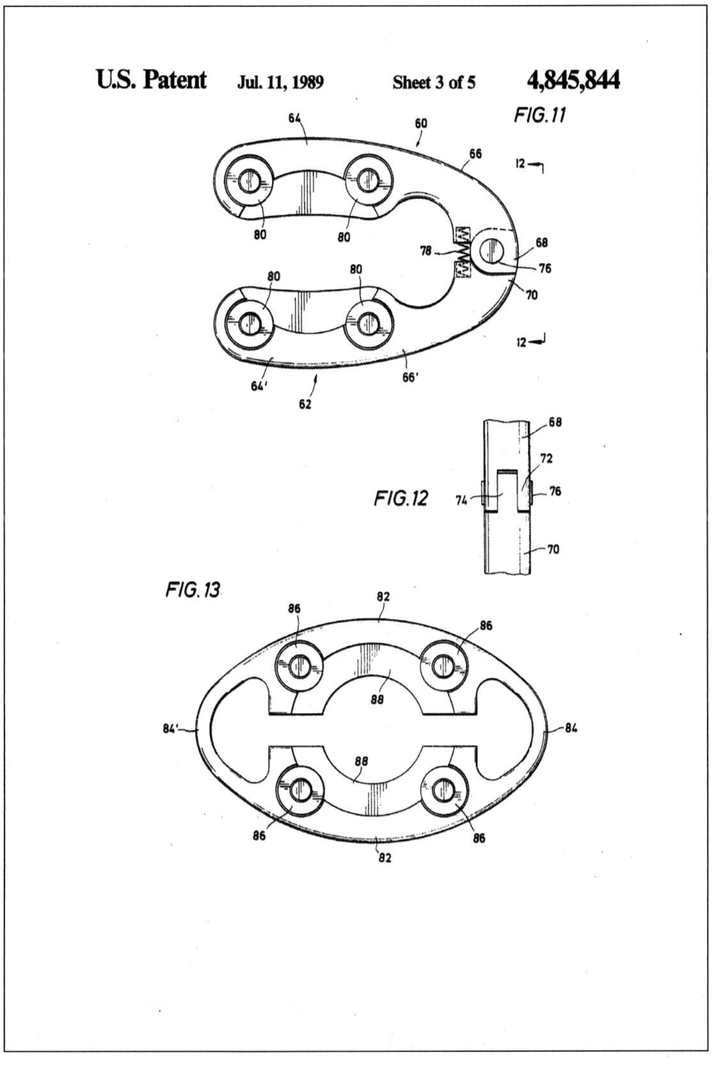
U.S. Patent Jul. 11, 1989 Sheet 3 of 5 4,845,844
FIG. 11
FIG. 12
FIG. 13

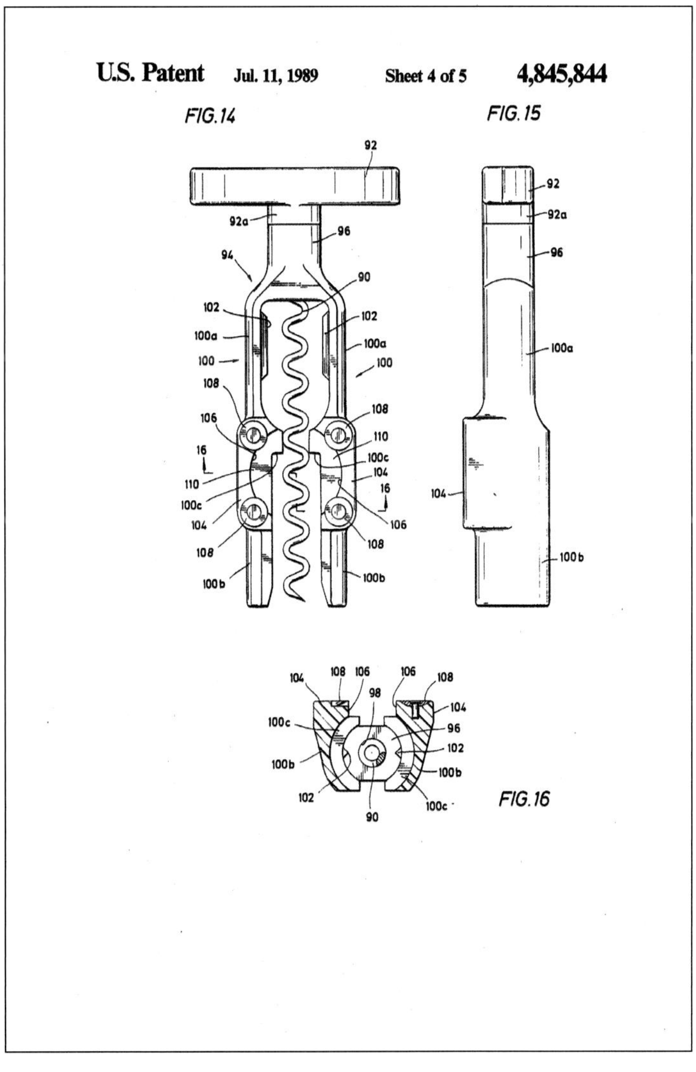
U.S. Patent Jul. 11, 1989 Sheet 4 of 5 4,845,844
FIG. 14
FIG. 15
FIG. 16

# United States Patent [19]		[11] **Patent Number:**	**5,056,676**
Allen, deceased et al.		[45] **Date of Patent:**	**Oct. 15, 1991**

[54] **BOTTLE CAP FOR REPEATABLE AIRTIGHT SEALING**

[76] Inventors: **Herbert Allen, deceased,** late of Houston; by Virginia D. Bedford, executrix, 5800 Lumberdale #64, Houston, both of Tex. 77092

[21] Appl. No.: **638,178**

[22] Filed: **Jan. 4, 1991**

[51] **Int. Cl.**[5] **B65D 45/24**

[52] **U.S. Cl.** **215/284;** 215/280; 215/287; 220/254; 220/233

[58] **Field of Search** 215/284, 287, 280, 293, 215/354; 220/254, 234, 233, 262, 90.4

[56] **References Cited**

U.S. PATENT DOCUMENTS

1,644,457	10/1927	Yancey	215/287
1,681,036	8/1928	Groby	215/293
2,038,489	4/1936	Guiot	215/284 X
2,170,531	8/1939	Kahn	215/280 X
2,529,817	11/1950	Russell	220/254 X
2,649,220	8/1953	Vastano et al. .	
2,966,277	12/1960	Bean	220/233
3,185,332	5/1965	Greenstein .	
3,317,071	5/1967	Teeter .	
4,009,794	3/1977	Zapp	215/210
4,190,173	2/1980	Mason et al.	220/203
4,567,994	2/1986	Hofmann	220/234
4,691,836	9/1987	Wassilieff	220/234 X
4,770,307	9/1988	Guglielmi .	
4,942,976	7/1990	Spencer	220/254

OTHER PUBLICATIONS

Device "D"—2 photographs of prior art device.
Device "E"—3 photographs of prior art device.
Device "F"—3 photographs of prior art device.
Device "B"—3 photographs of prior art device.
Device "c"—2 photographs of prior art device..

Primary Examiner—Stephen Marcus
Assistant Examiner—Vanessa M. Roberts
Attorney, Agent, or Firm—Browning, Bushman, Anderson & Brookhart

[57] **ABSTRACT**

An improved bottle cap is provided for repeatedly obtaining airtight sealing engagement with a bottle having an annular external lip. The cap of the present invention is particularly well suited for sealing bottles containing sparkling wine or champagne, since such bottles conventionally include an external lip and since the cap of the present invention is able to seal gases within the bottle. The cap includes a housing having a central aperture for positioning over the neck of the bottle and having a radially inward directed flange for fitting beneath the external lip on the bottle. An activating carrier is movable within the aperture in the housing, and a ring-shaped pad holder and elastomeric pad are mounted on and are axially movable with respect to the carrier. A biasing spring is provided for acting on the holder and the pad to bias the pad toward sealing engagement with the bottle. A handle is rotatably secured to the housing, and includes a cam lobe for engaging the carrier and thereby energizing the spring to force the pad into sealing engagement with the bottle. An improved method is provided for fabricating a bottle cap for sealing engagement with a bottle having an annular external lip.

20 Claims, 1 Drawing Sheet

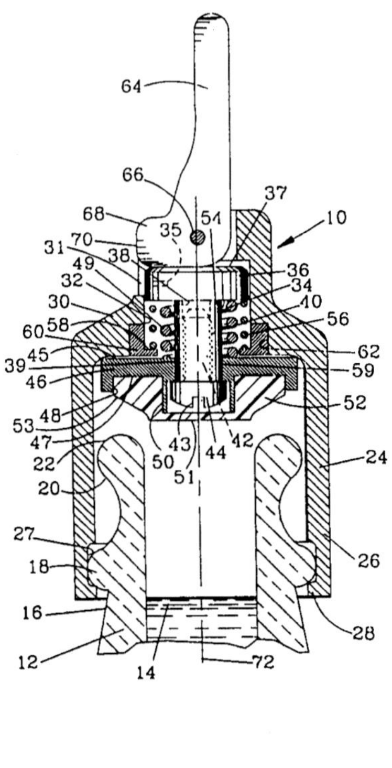

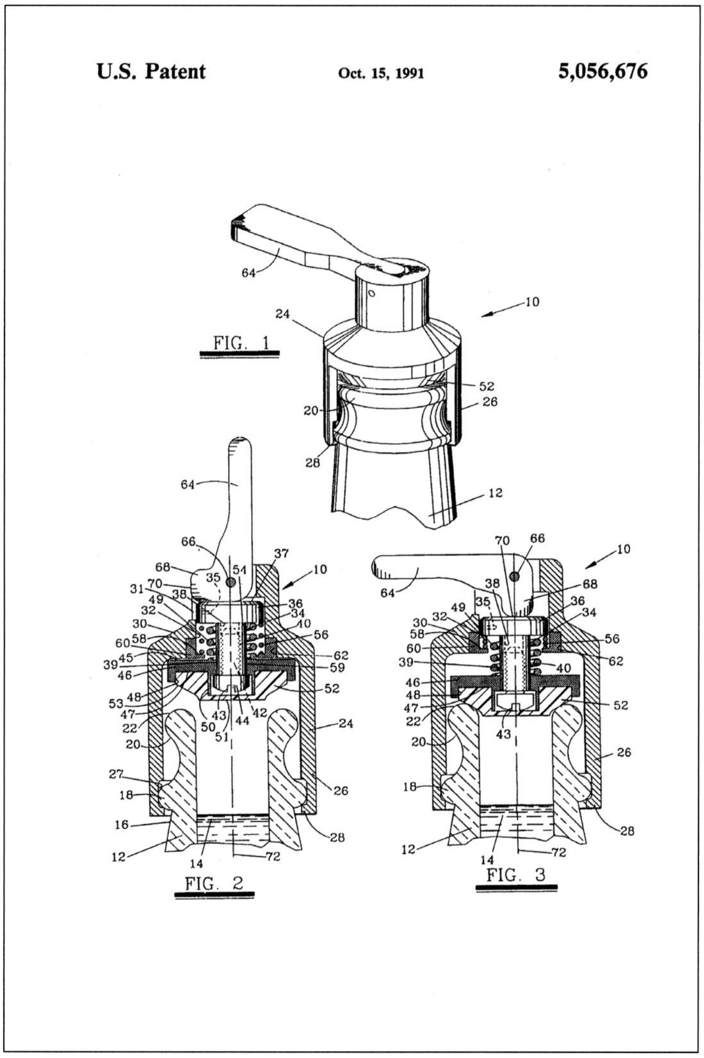

Note: Seperate drawings for the rack-and-pinion style Lever Pull were never submitted for patent. They were manufactured under the patent of the lever style Lever Pull. This is apparently not unusual in the world of patents, where a product is manufactured under a patent with drawings that are entirely different from the final product.

Screwpull®

Chapter 12: Packaging and Marketing

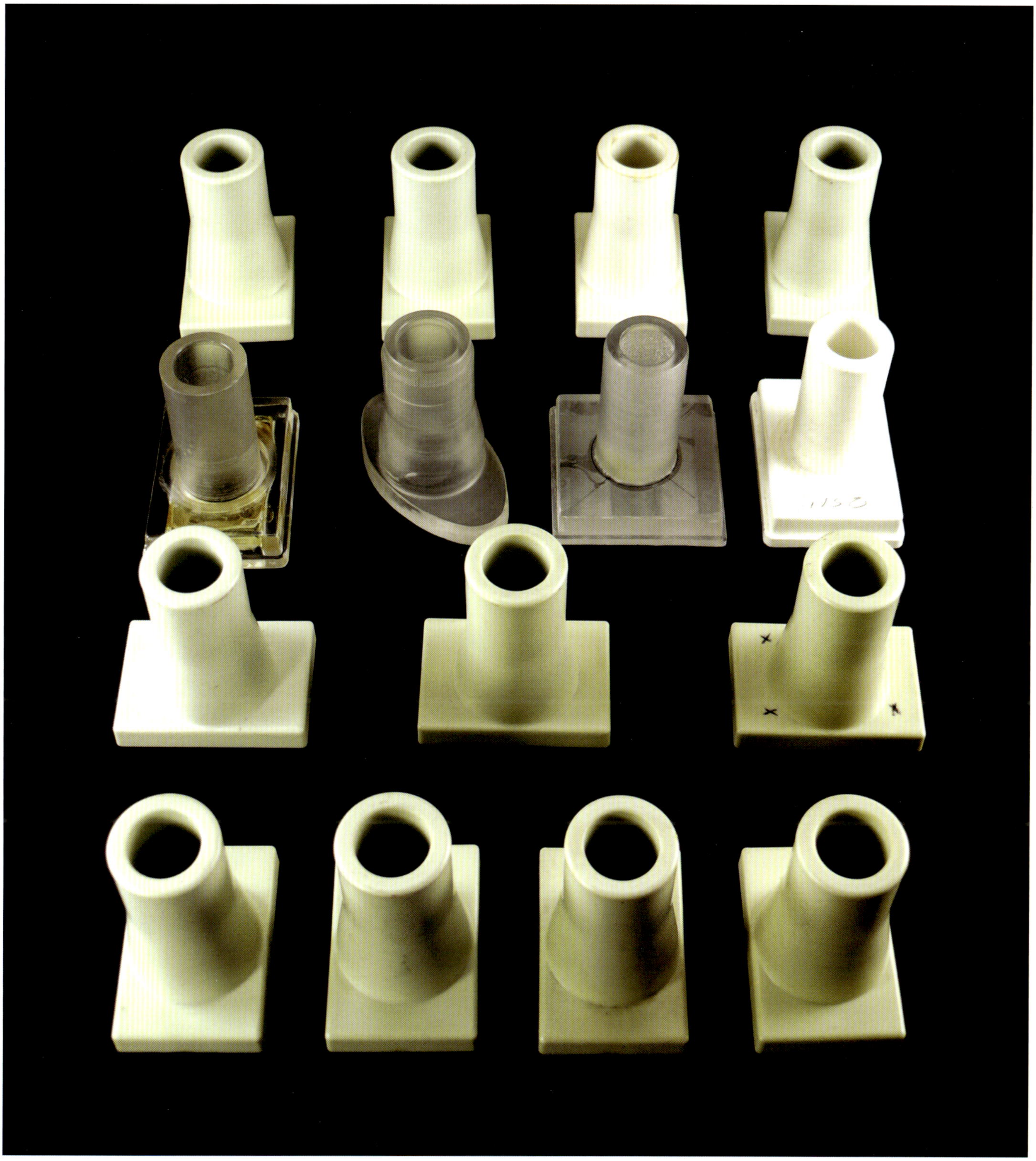

Prototype of Table Model case and base.

Production Model of Table Model case and base.

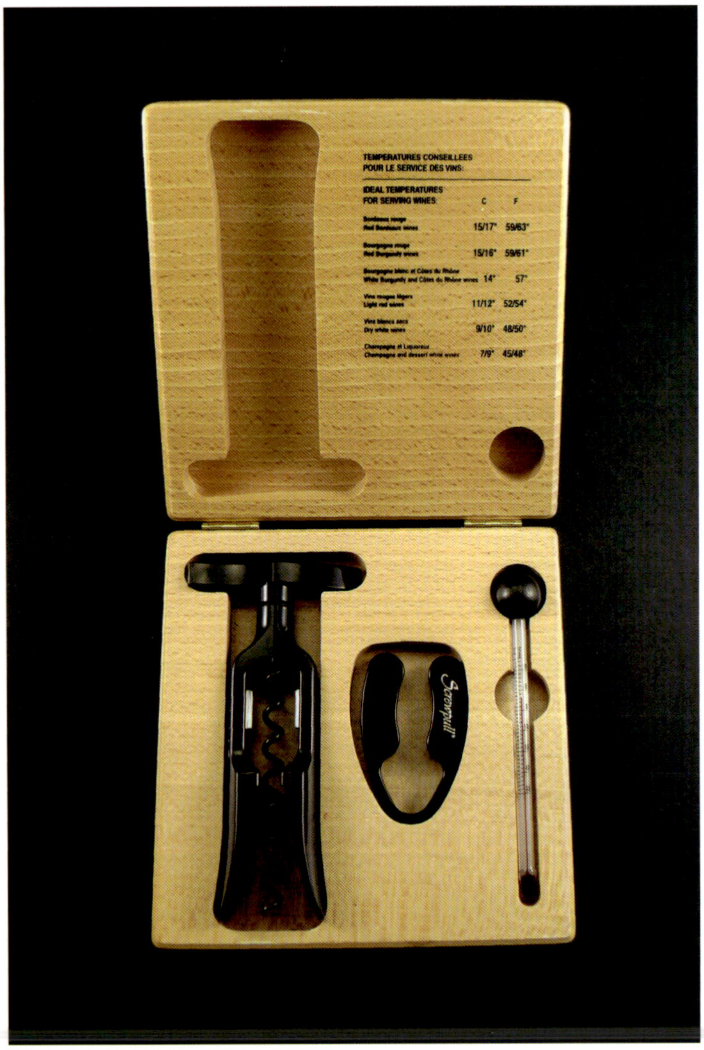

Screwpull Table Model with Foil Cutter and thermometer with ideal wine temperatures shown on the inside of the lid, packaged in a light-oak box. *LeCreuset.*

Advertising and instruction cards and/or pamphlets included in Screwpull cases. (above and next two pages)

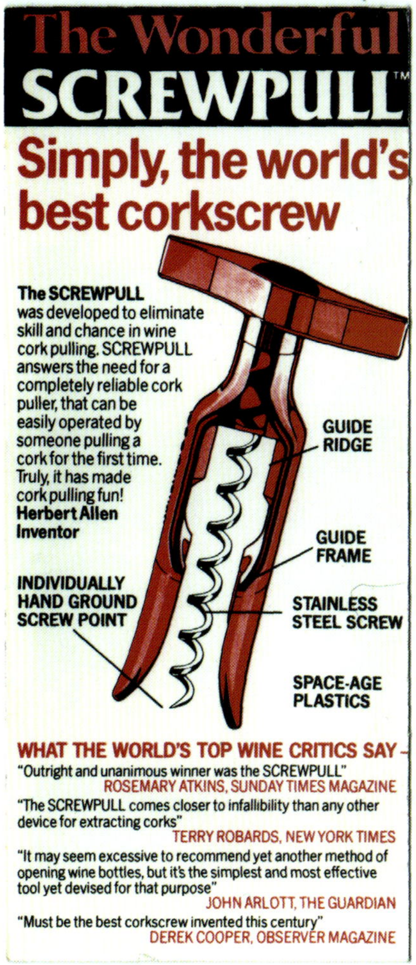
The Wonderful
SCREWPULL™
Simply, the world's best corkscrew
The SCREWPULL was developed to eliminate skill and chance in wine cork pulling. SCREWPULL answers the need for a completely reliable cork puller, that can be easily operated by someone pulling a cork for the first time. Truly, it has made cork pulling fun!
Herbert Allen
Inventor
GUIDE RIDGE
GUIDE FRAME
INDIVIDUALLY HAND GROUND SCREW POINT
STAINLESS STEEL SCREW
SPACE-AGE PLASTICS
WHAT THE WORLD'S TOP WINE CRITICS SAY
"Outright and unanimous winner was the SCREWPULL"
ROSEMARY ATKINS, SUNDAY TIMES MAGAZINE
"The SCREWPULL comes closer to infallibility than any other device for extracting corks"
TERRY ROBARDS, NEW YORK TIMES
"It may seem excessive to recommend yet another method of opening wine bottles, but it's the simplest and most effective tool yet devised for that purpose"
JOHN ARLOTT, THE GUARDIAN
"Must be the best corkscrew invented this century"
DEREK COOPER, OBSERVER MAGAZINE

Screwpull™
WINE ACCESSORIES BY HALLEN
Screwpull with Foil Cutter Gift Set
The SCREWPULL is a foolproof corkscrew that puts an end to any struggling with wine bottles. All you have to do is twist the handle at the top. The stainless steel screw enters the cork, grips and removes it, all in one, failsafe action. No broken or crumbled corks, and no pulling. The FOIL CUTTER is the only invention that tackles the problem of removing the foil cap from wine bottles. With a simple twist the Screwpull Foil Cutter will neatly cut through any foil or even the tougher plastic capsules.
NOW AS A SET
Acclaimed by wine experts the world over, The Screwpull with Foil Cutter will make a perfect and lasting gift for anyone who enjoys wine.
ictc
2 Fleming Way Isleworth TW7 6EU

Screwpull™
GIFT SET BY HALLEN
New!
Screwpull
and
Foil
Cutter
Gift
Set

SCREWPULL T.M.
Space Age Plastics
Super strong yet flexible enough to spread over the larger bottles—even a Nebuchadnezzer.
Guide Ridge
Prevents rotation of cork-thereby insuring its complete removal from the bottle as you turn the handle—no pulling required.
Guide Frame
Directs the screw toward the center of the cork.
Stainless Steel Screw
An anti-friction coating reduces to an absolute minimum the effort required to turn the handle.
Swing-Out Knife
Handy for removing bottle seals and tops.
Individually Hand Ground Screw Point
Carefully ground to a special shape that prevents knocking off fragments as it penetrates the bottom of the cork.
A product of the Hallen Company
P.O. Box 1392
Houston, Texas 77001
Telephone:
Toll Free 1-800-231-2746
In Texas (713) 683-8111
Hallen Company International
#6 Hays Mews
London WIX 7PR
England
Telephone:
499-8331
Patented and Patents Pending

The Pocket
SpinHandle
NEW!
* Opens wine bottles with only one finger — in a flash!
* Now with built-in knife/bottle opener!
Screwpull
WINE ACCESSORIES BY HALLEN

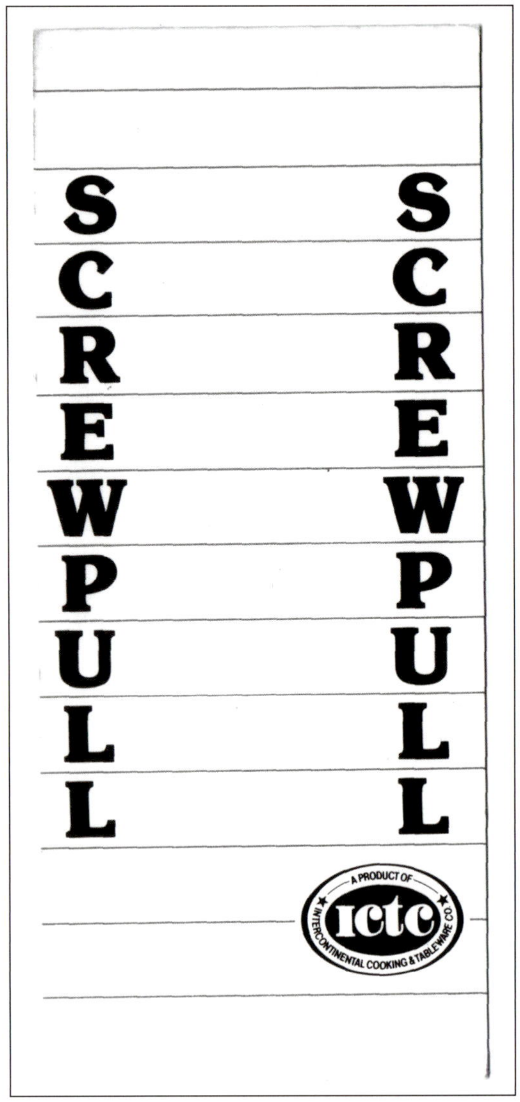
SCREWPULL
SCREWPULL
A PRODUCT OF
ICTC
INTERCONTINENTAL COOKING & TABLEWARE CO.

Screwpull®
by LE CREUSET

The side of the box of the first Leverpulls. Note the gold finish on the logo. Leverpull #1 had its logo colored gold, I am sure for this photo. That is the only place I have seen the gold on a Leverpull ad and #1 is the only place I have seen it on an actual Leverpull.

8.5 x 11 mailer and advertisements, c.1982.

Later Leverpull boxes. *Courtesy of Karel Janousek.*

8.5 x 11 mailer and advertisements, c.1982.

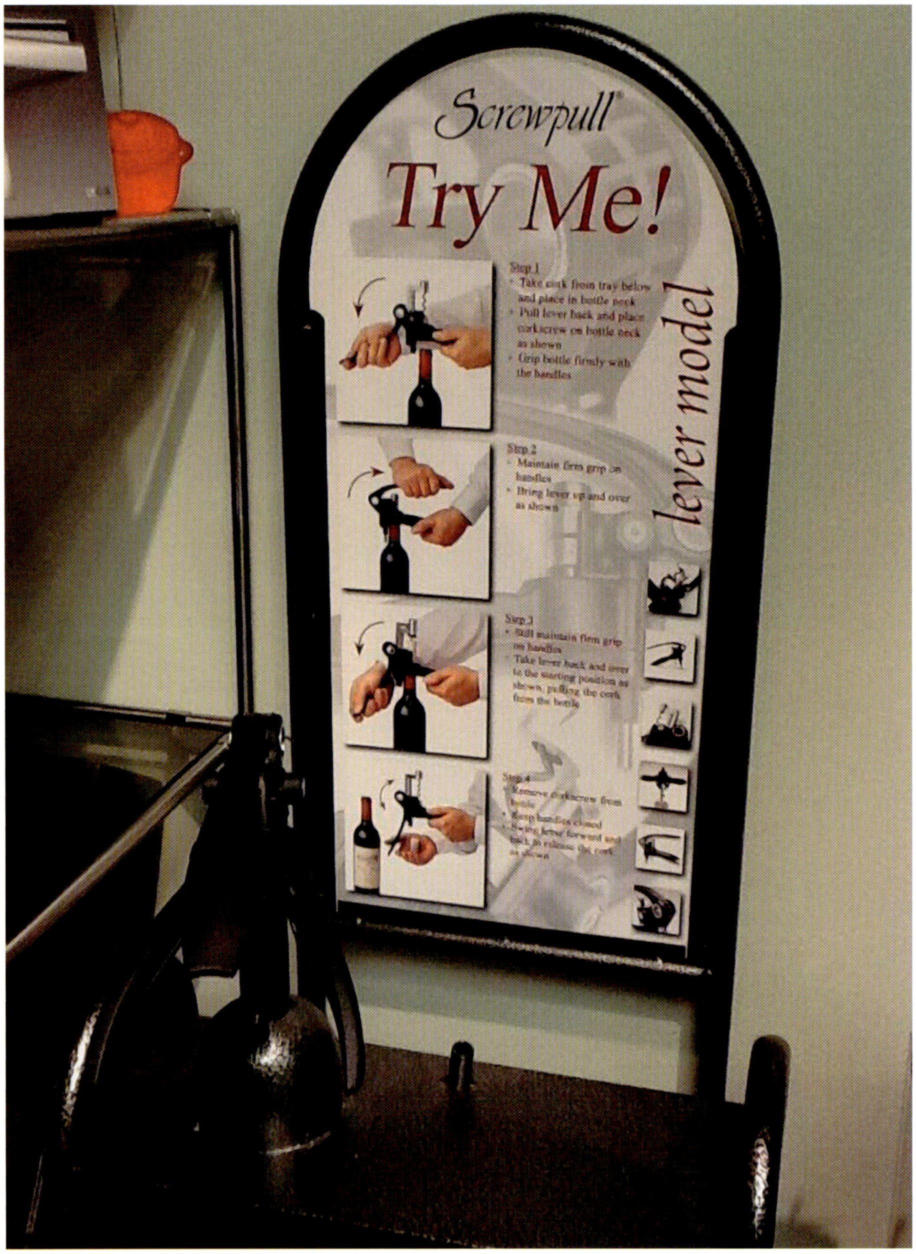

A LeCreuset "Try Me!" display used at retail stores where you can actually try using the Leverpull.

Ads such as this one, which ran in the *Houston Home & Garden Magazine*, November 1986, were placed in many key magazines and periodicals from the early 1980s through the 1990s.

8.5 x 11 flyer, c.1984.

From that time to present, the family of Screwpull products has been improved, increased, and manufactured by LeCreuset of France. LeCreuset made an improvement to the Lever Models, in 2000. Previous models incorporated an axle pin, for the pull lever to pivot on, of approximately 1/10" in diameter. The new models have an axle pin of 7/8" in diameter. They made the pin with a 5/8" open hole in the center, whether for weight reduction or just design purposes, but it somehow increases the mechanical advantage of the machine. The "feel" is much better and seems to take less effort to pull the lever while removing the cork. Most of the newer models of the Screwpull lever corkscrews incorporate this new axle pin design.

After LeCreuset took control of manufacturing the Screwpull line, they continued making the Version 3 Table Model with the same colors as produced by the Hallen Company, but eventually increased their color line to include many other fashionable colors. They also continued making the Original Leverpull for some time before improving it and adding new designs to the line.

Table Models in some of the new colors. *Courtesy of Don Bull.*

Table Model, design by LeCreuset, called the "Elite," with a foil cutter built into the legs of the lower frame.

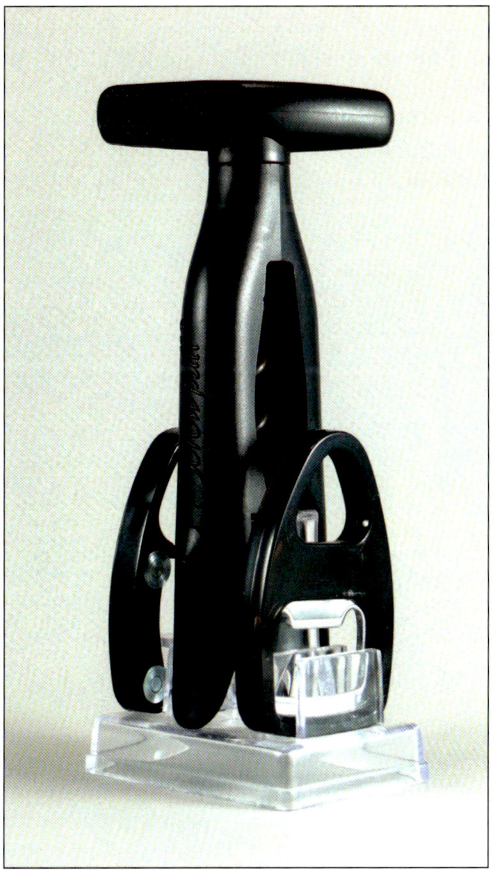

The Elite with a separate Foil Cutter and a separate Cap Lifter in a Gift Set.

These two appear to be knockoffs of the Elite. The plastic material, the short screw, and the sloppy fit at the top does not seem to be the way Screwpull puts a product on the market. They are unmarked.

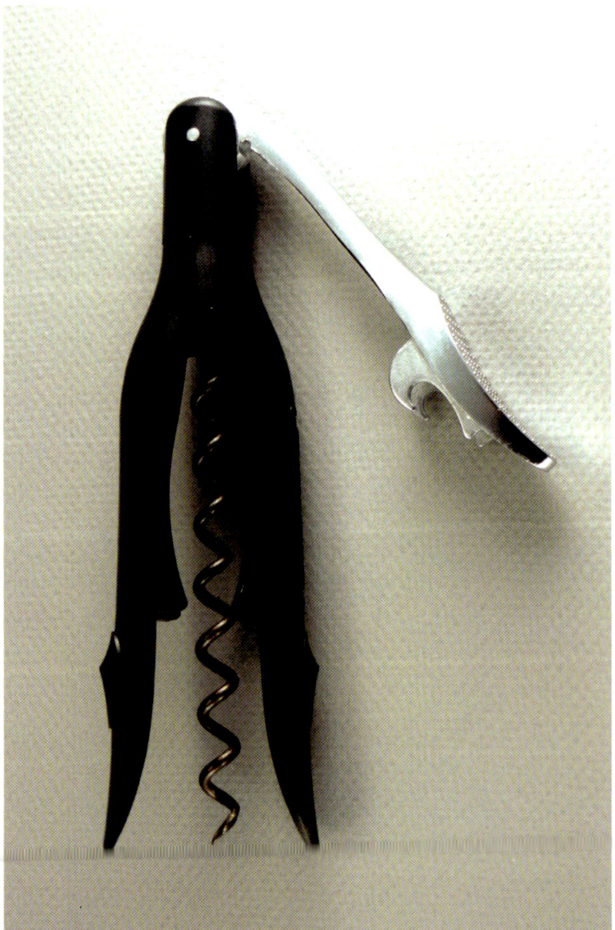

A Spin Handle type Table Model, called the "Bistro" with built-in foil cutter in the lower legs of the frame and cap lifter on the end of the handle. The Bistro uses the same frame or body as the Elite.

A LeCreuset Table Model without the built-in foil cutter in the lower legs of the frame, The Classic TM110 is sometimes called "The original table model Screwpull." The letters of the Screwpull logo are indented into the lower half of one of the legs instead of raised letters in the upper half of one of the legs as in the "original" Screwpulls. Made in China.

Triology Table Model with Spin Handle, Red Ball, and Foil Cutter.

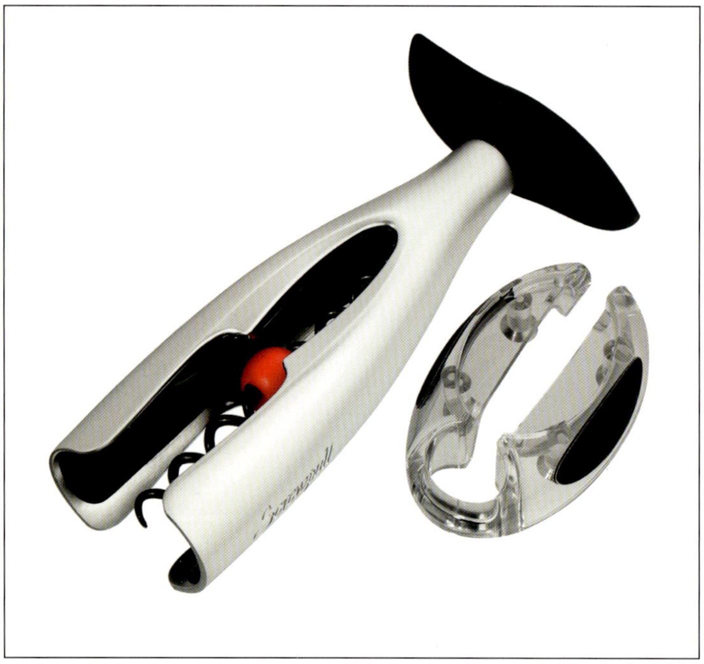

Triology Table Model with Foil Cutter. The Activ Red Ball in the Triology acts much like the tube in Herb Allen's cork ejector patent. By reversing the turning action of the screw, the Red Ball pushes the cork off of the screw. See chapter 8 for the cork ejector patent.

New style Foil Cutter.

Foil cutter included with the new LM G10.

Three-in-one Pocket Model with Spin Handle, cutters, and openers.

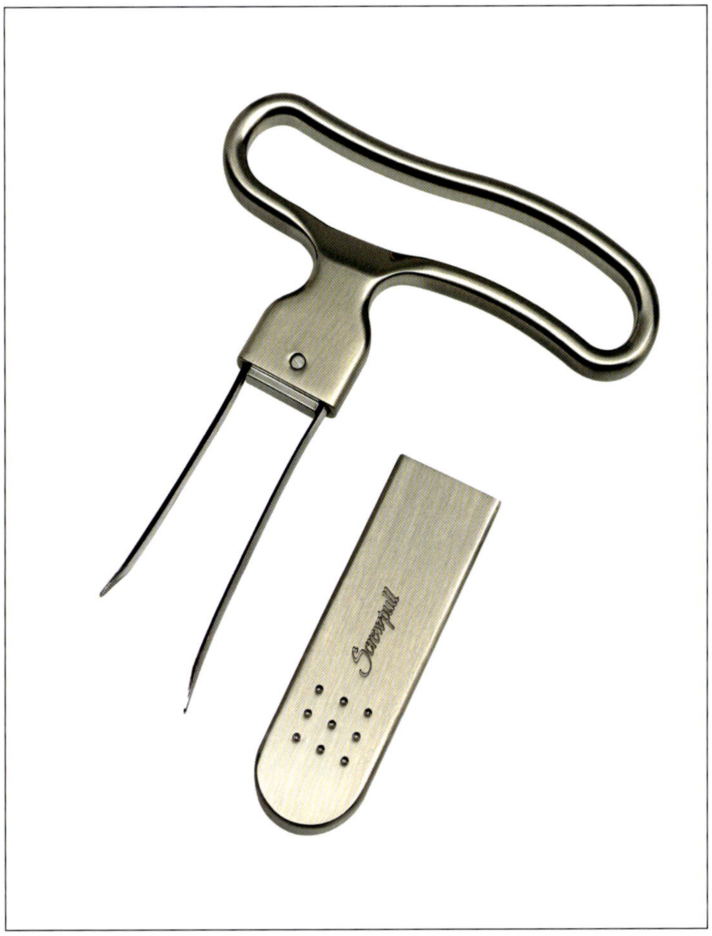

Butler's Friend, a two-pronged opener. *Courtesy of LeCreuset.*

Waiter's Pocket Screw. *Courtesy of LeCreuset.*

Screwpull Leverpull Production Model, #0297, made in France.

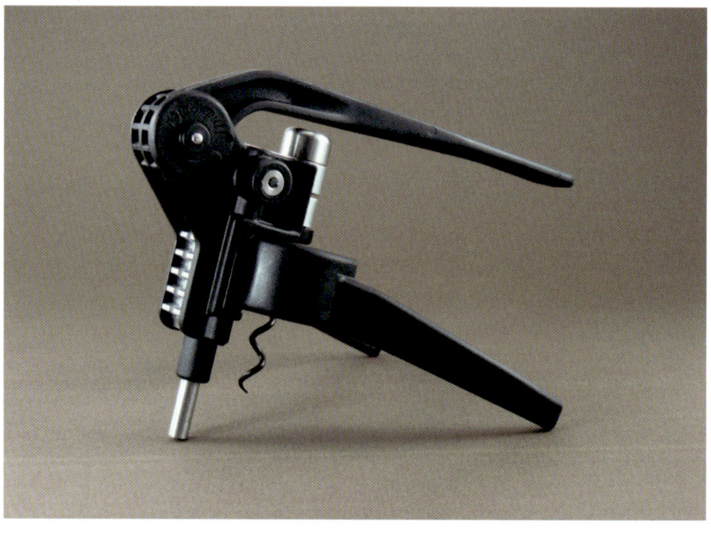

Screwpull Leverpull Production Model, #0304, made in France.

Leverpull Production Model #0899, made in France. Logo at side of corkscrew does not have the "Patent Pend" part of the Screwpull logo. An early LeCreuset Production Model, it is called "Professionele."

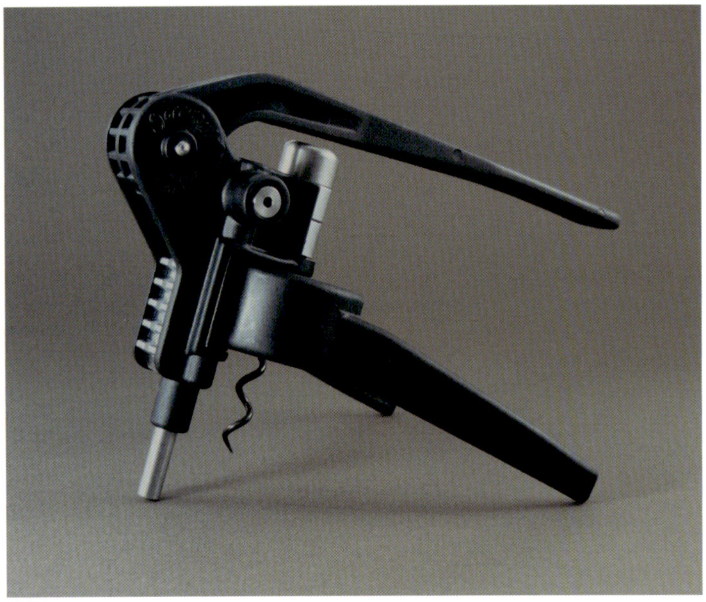

Screwpull Leverpull Production Model, 39636, made in France.

Screwpull Leverpull Production Model in Tortoise Shell finish, no numbers, made in Ireland.

Screwpull Leverpull Production Model in a checkered pattern finish, no numbers. Date stamped through directions booklet, April 28, 1998, assuming date of manufacture. Made in Ireland.

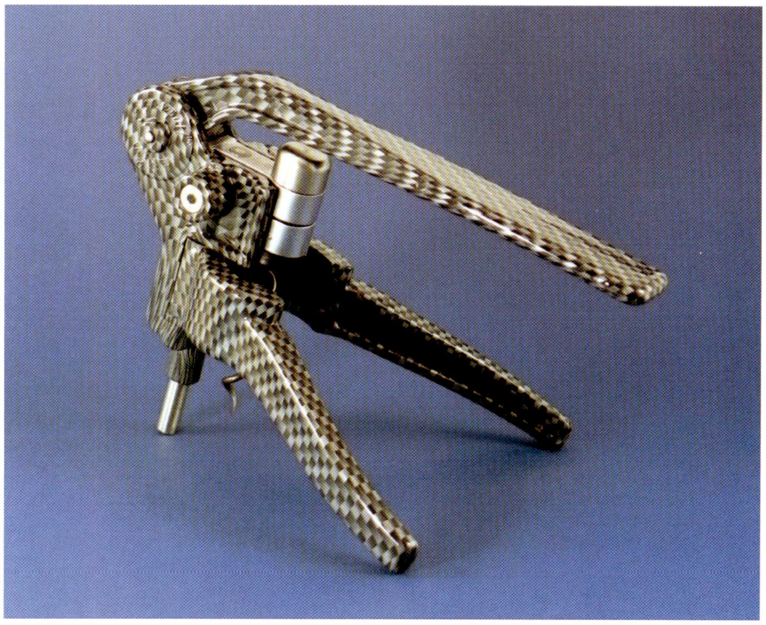

The LX Professional Production Model, LM200, #B014486, c. 2001, the first real Leverpull design change. Note the larger open-hole, lever handle axle-pin design.

The Elegance, Leverpull Production Model #005, made in Ireland, c. 2000. Note the smaller lever handle axle pin.

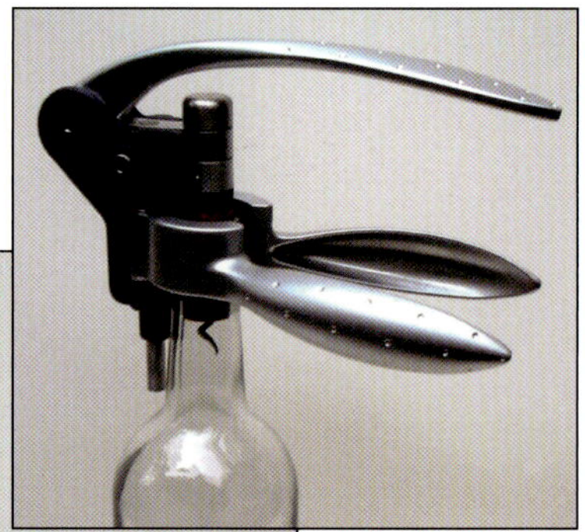

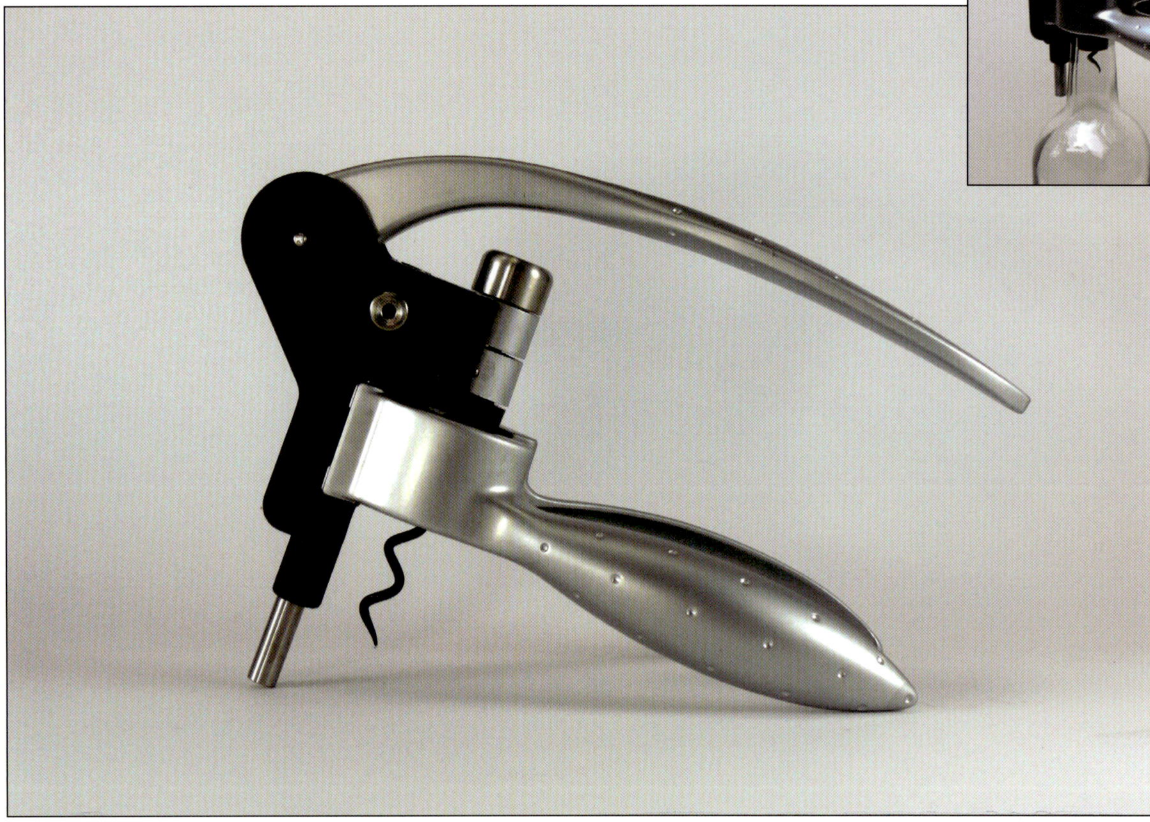

The Stealth, Leverpull Production Model #B309026, made in Ireland, c. 2003.

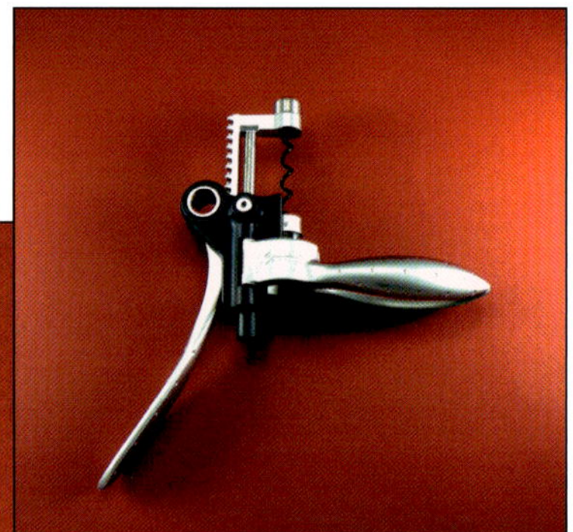

The LM-300 Leverpull Production Model #C850929, made in Ireland, c. 2005. The last Screwpull leverpull type corkscrew to have serial numbers.

A Trigger Pull prototype. It appears to be made of a plastic composite material painted a silver color. The trigger is made of a white plastic. Note the shape of the front over the rack, it is almost a straight line from the top down where the production models go down about 45 degrees and then curves to straight down.

Trigger Pull prototype. It is made of a black composite material with the same white plastic trigger. Note that there is no logo on the side of the corkscrew. Prototypes rarely had the logo incorporated into their construction.

Another Trigger Pull prototype made of black composite material with the same white plastic trigger. Note the shape of the front, over the rack, has the curve in the center but has a slight curve back near the bottom, instead of straight down like the final production model. It does not have the raised Screwpull logo on the side.

The Trigger Pull Leverpull, LM350, Production Model version in composite material. Note the logo. Serial numbers are not found on any of the Leverpull models from this point on. The LM300 was the last model to have serial numbers.

Wine Enthusiast version of the Trigger Pull. Note the black "hubcap" over the open hole lever pin axle and the scalloped finger curves in the trigger. There is no raised logo on the side, perhaps a knockoff.

The Trigger Pull Leverpull, LM350B, Production Model in satin chrome metal. All parts are metal.

The Trigger Pull, LM350B in black nickel finish.

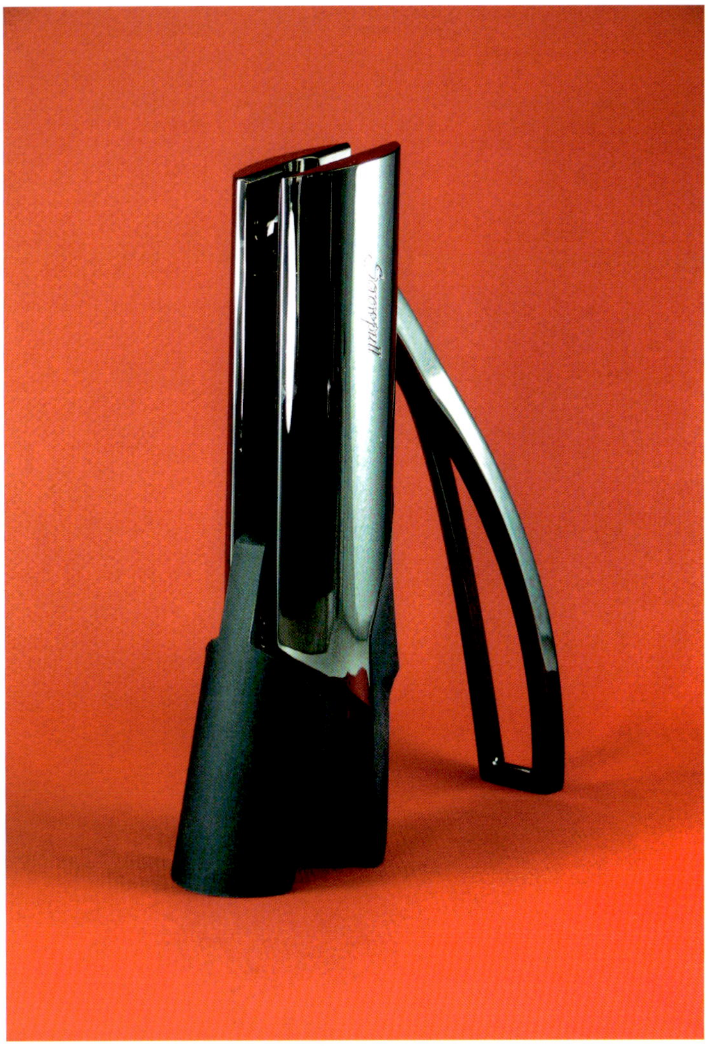

The LM400 Vertical style Leverpull Production Model. A beautiful design, heavy, and built like a tank.

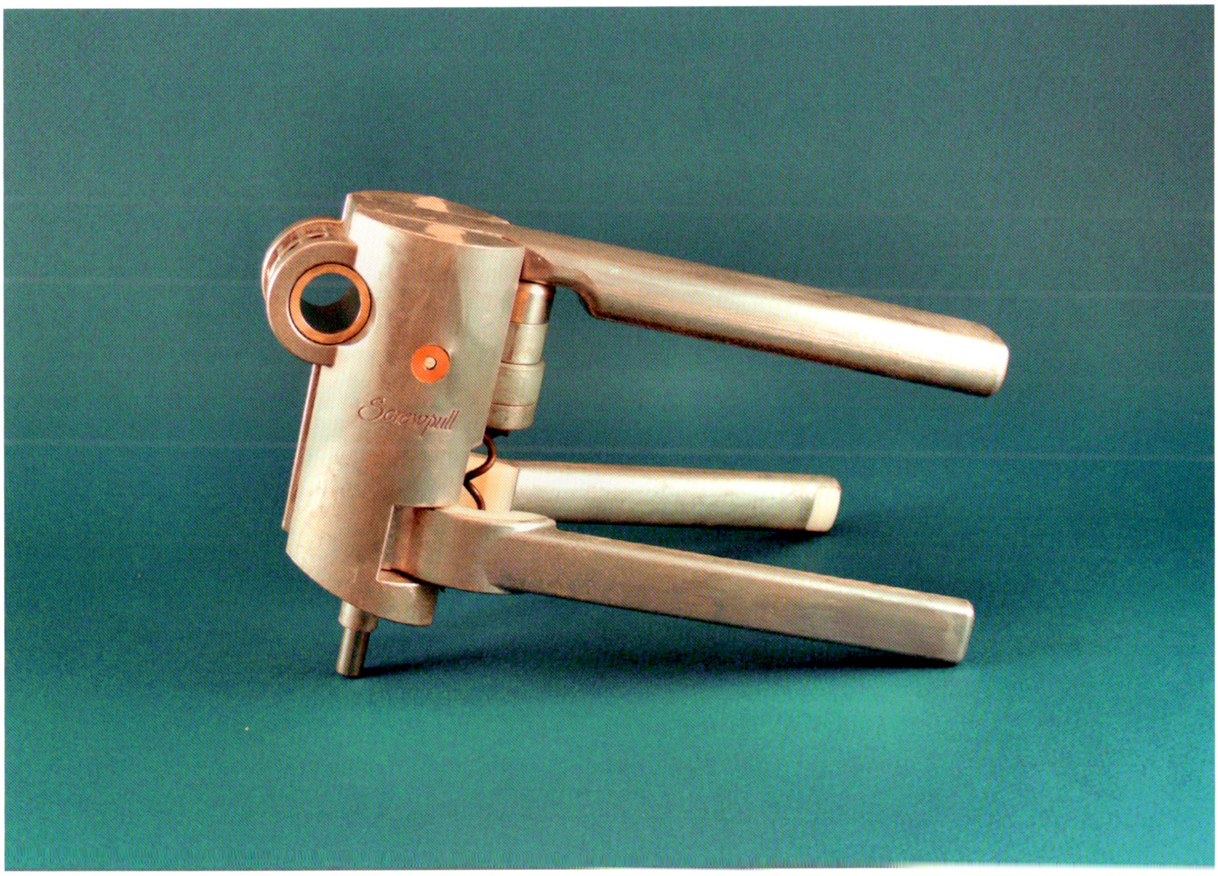

LM G10 heavy duty Advanced Lever Leverpull prototype.

The LM G10 prototype with no finish on the metal.

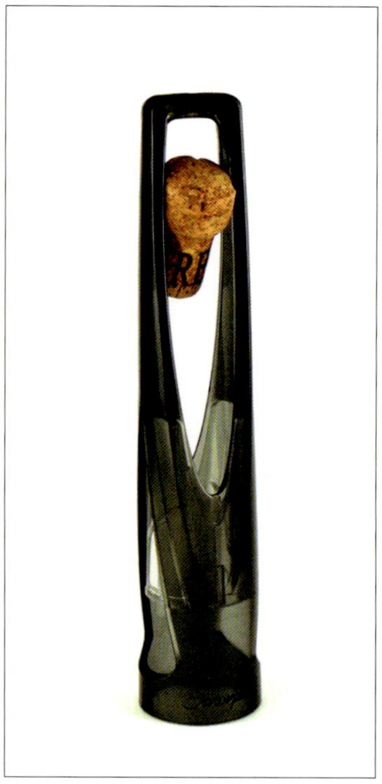

The Cork Catcher in the plastic version. It is also made in a metal.

The LM G10 Advanced Lever Production Model with black nickel finish. Note logo name compared to the previous LM G10. The Screwpull logo will no longer exist in the U.S., Canada, and many other parts of the world.

The LM G10 Advanced Lever Production Model with bright and satin nickel finish. Take a good look at this corkscrew, you may never see another on the U.S. market with the Screwpull logo on it. See the logo in the next photo.

The all-metal, black nickel satin and bright chrome-finished, "Winged" double-lever corkscrew.

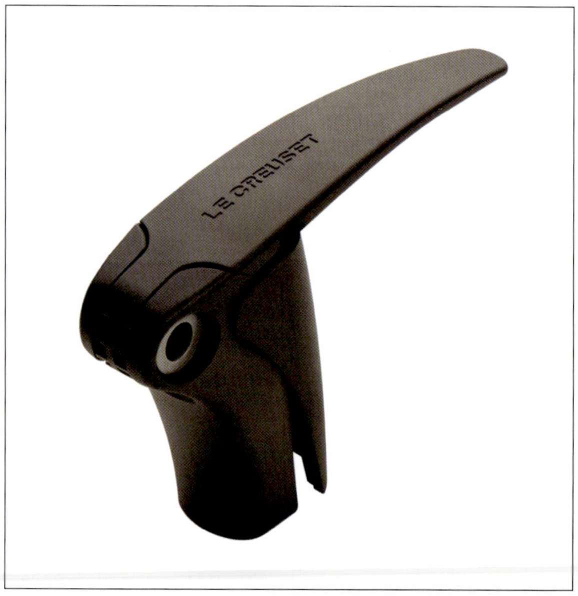

The newest lever, LM150. Note the logo.
Courtesy of LeCreuset

The LM G10 Advanced Lever.

Chapter 14: The Patent Expires

In just a few short years after the March 3, 1981 Leverpull patent expired, the market exploded with reproduction versions of the Leverpull. Most of the knockoffs are made in China for American companies to be sold for a price much less than the Leverpull.

Several versions of knockoffs are constructed nearly equaling the quality of the Leverpull, but there are some not worth the price. Some knockoffs are selling for less than the cost of a Screwpull Table Model.

It is obvious that there are several models made by the same manufacturer in China with only logo changes for identical models and/or different pull-lever and grip-handle designs for differently named models. Most of the working parts are interchangeable. The reproduction models do not even bother to vary the mechanisms. Prolonged testing is unnecessary since they are copying a proven product. The only clue to where most of the units are manufactured is a small, gold, oval sticker: "Made in China." If the units are packaged in a box, the box sometimes has a note printed on it: "Made in China." Herb Allen considered the copying of his products a great compliment, but once stated that it sure does eat into your market.

There are at least forty different lever model knockoffs in circulation—many with different color and material selections. Additionally, there are knockoffs of the vertical type where your hand holds the bottle grips around the neck of the bottle instead of off to the side, like the Leverpull. These have similar, but different, mechanisms and work quite well for removing real corks. They are a complete flop on harder, more dense corks.

Most lever-model corkscrews, including the Screwpull Leverpulls, have warnings in their instructions such as: "Not recommended for rubber or plastic corks," or "Do not use on dense or composite corks," or "Some synthetic corks can bend or damage the corkscrew worm," etc. Using these corkscrews on hard, dense, composite, or synthetic corks is a problem. You cannot pull the cork or easily remove the screw. The screw will just not let go of the cork. It takes some doing and mental anguish to unscrew the screw back out of the cork. Finally, you get the screw out and go for your trusty Screwpull Table Model to get the job done. Even with the warnings, how can you easily deduct that the cork is a hard composite with real cork ends? You don't, until the screw is part way into the cork and all you can say is, "Aww s—."

More and more wineries are using hard composite and/or synthetic corks, and even (heaven forbid) screw caps. As fewer real corks are used, the market for lever-model corkscrews will dwindle to those customers who maintain cellars of great wines and still use real cork seals. Direct pressure insertion becomes almost impossible on the very dense corks, even with Teflon coatings to reduce the friction. There comes a point where, if you designed a longer lever, greater mechanical advantage gears, longer pitch screw, etc. you would either push the cork into the bottle or bend or break the screw or all of the above.

The Leverpull definitely is the greatest contribution to mechanical corkscrew design in the twentieth century, but even the Leverpull has its limitations.

Chapter 15: Leverpull Knockoffs

The photo on the previous page is a moonrise over the White Fences Winery and vineyards, c. 2004, showing a giant corkscrew at the entry. The photo is an "Artography" composite by professional photographer Michael Neff. There actually are two corkscrews, one on each side of the entrance drive. I would venture to say that these are the two largest Screwpull knockoffs in the world today.

This chapter contains nothing more than way too many knockoffs of the Leverpull type corkscrew. I am showing you this many of them so that you can comprehend how one product, such as Herb Allen's Leverpull generated much interest in the business world, around the world. The following photographs represent many different combinations of designs, colors, packaging, and quality of the copies manufactured. Most knockoffs come with a foil cutter and an extra screw. Many have other items, such as stands, aerators, stoppers, and even two extra screws. The packaging varies from being shrink-wrapped on a cardboard backing to beautiful, exotic wood boxes lined in cloth and wood boxes covered in leather. To see many other knockoffs that exist, go to "corkscrews" on eBay.

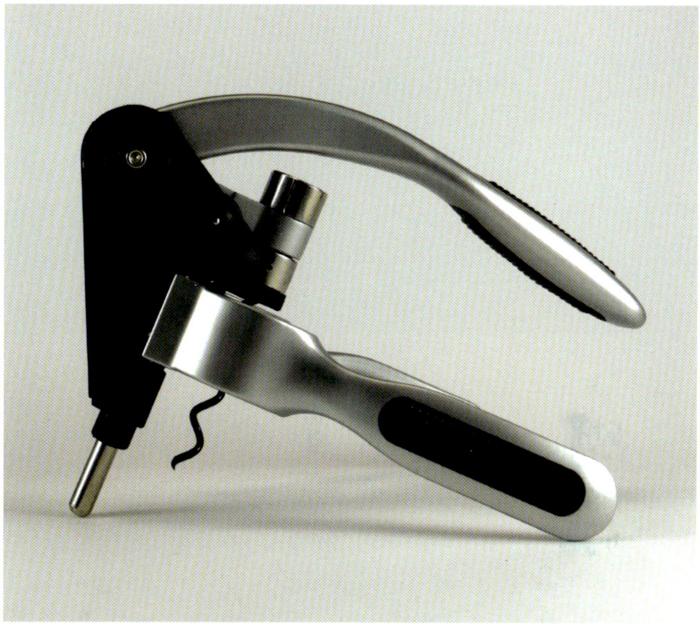

Wine Mastique.

From The Sharper Image #P1500, made in China.

The VinoPull Uncorker, distributed by Franmara, Inc. also marketed as the Cork Genie, made in China, 2001.

Pull It, made in China

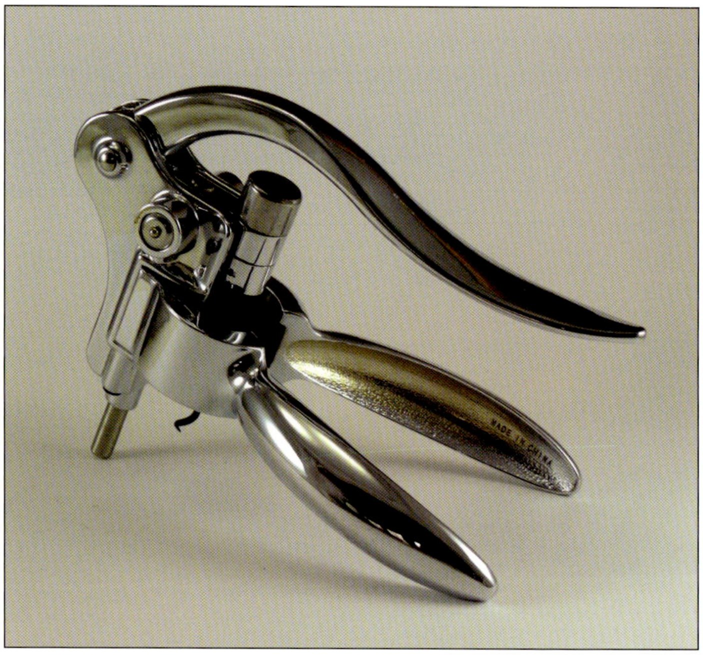

Wine Bottle Opener, by OGGI Corp., made in China.

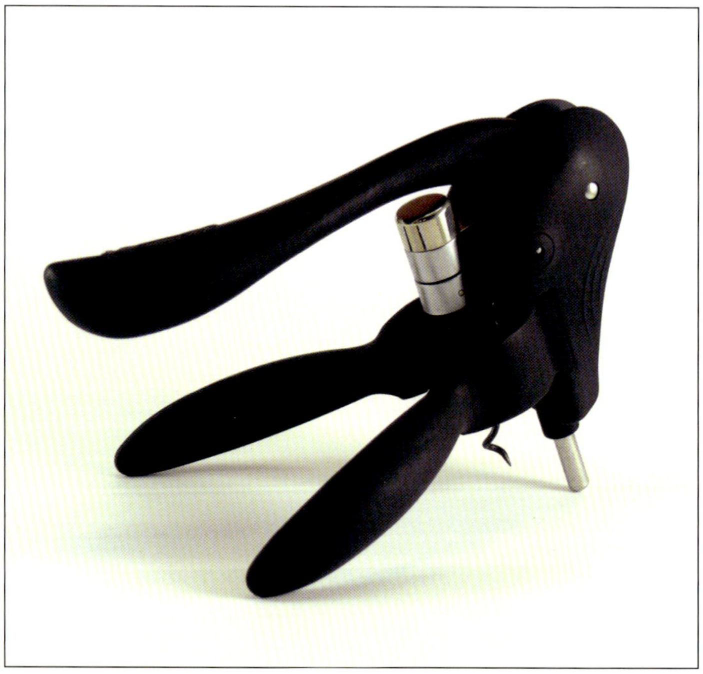

EZ Pull by Farberware® Inc., made in China.

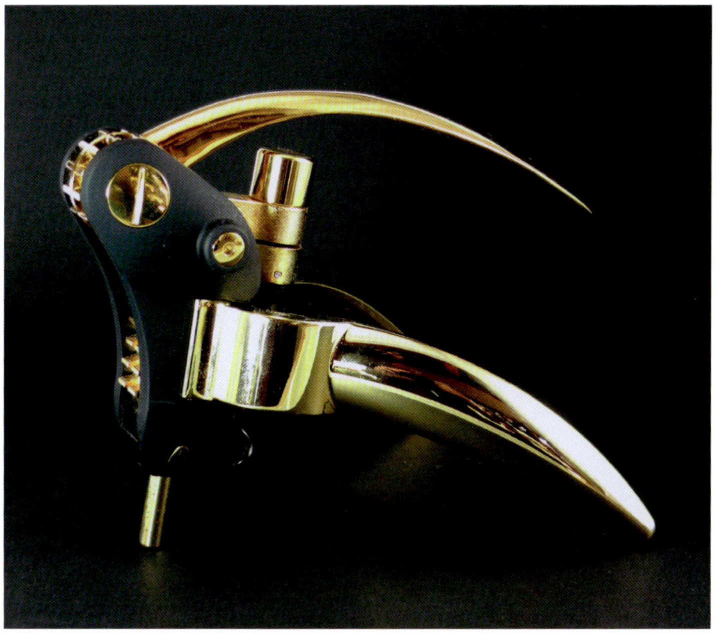

Pull It Gold Edition, genuine 24-carat gold plated finish, made in China.

No name, very light weight metal.

Easy UP™, made in China.

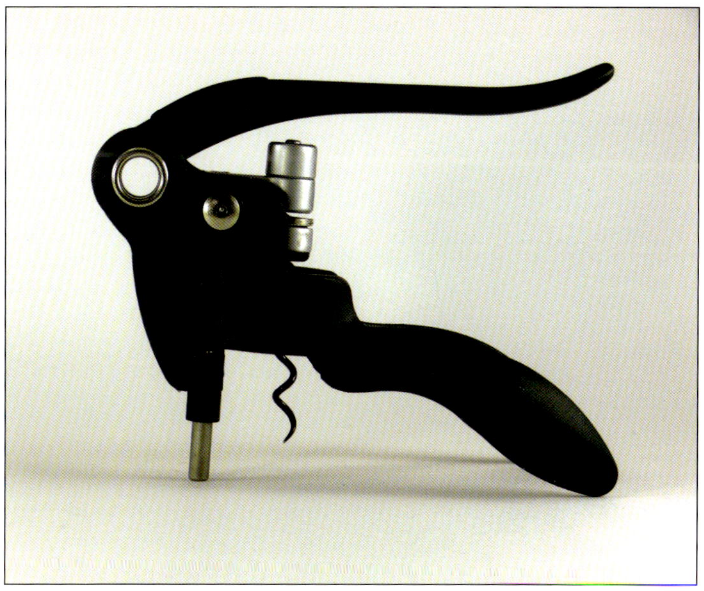

Corkscrew, Deluxe Corkscrew.

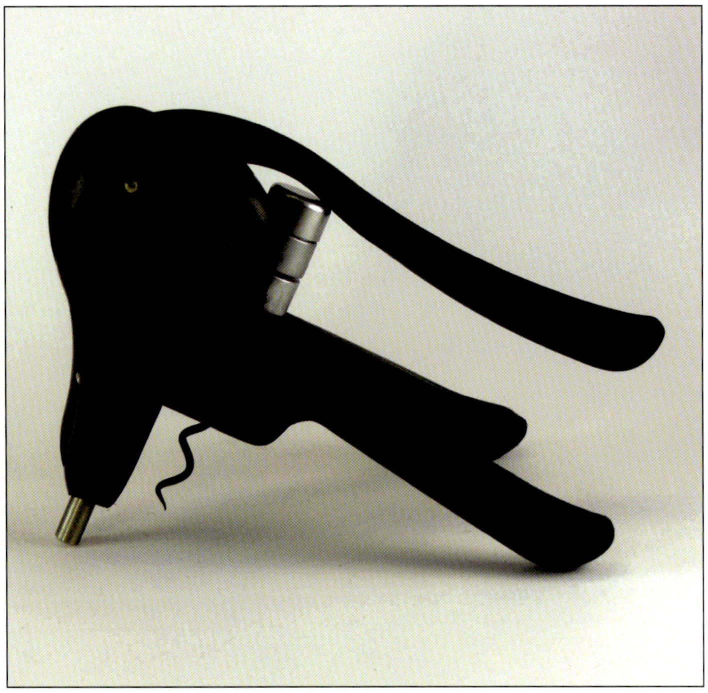

Deluxe Corkscrew, by Norpro, made in Taiwan.

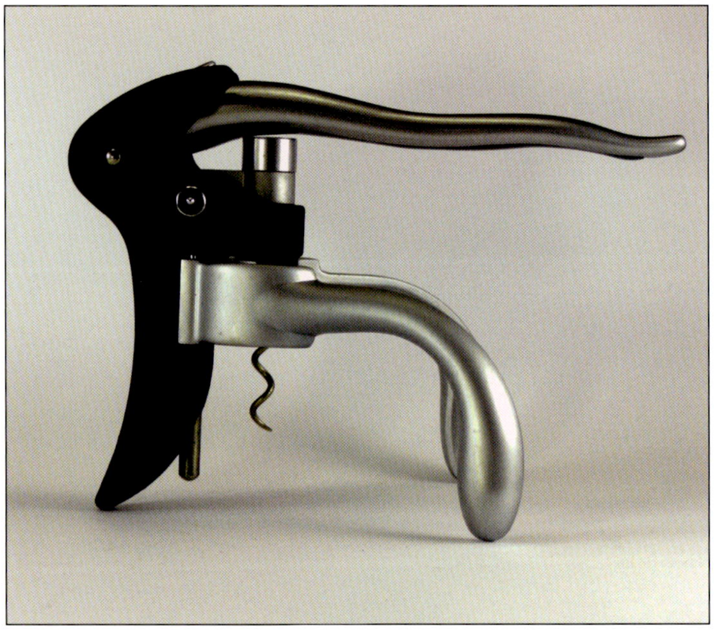

Cuisena, by Fox Run Craftsmen, made in China.

Quicksilver Deluxe Corkscrew, by *Wine Enthusiast*, made in China.

Arcosteel heavy metal also marketed as Easy Lift by Bed Bath and Beyond, made in China, 2001.

ColorPull by Oenophilia.

Professional Corkscrew, by Costco Corp., made in China.

Silver Spur black plastic lever corkscrew.

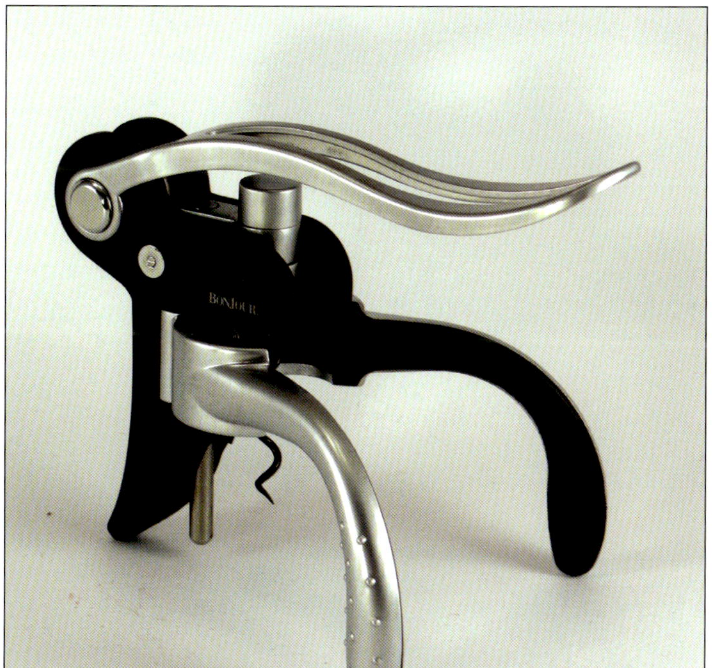

BonJour Chateau Deluxe corkscrew, made in USA.

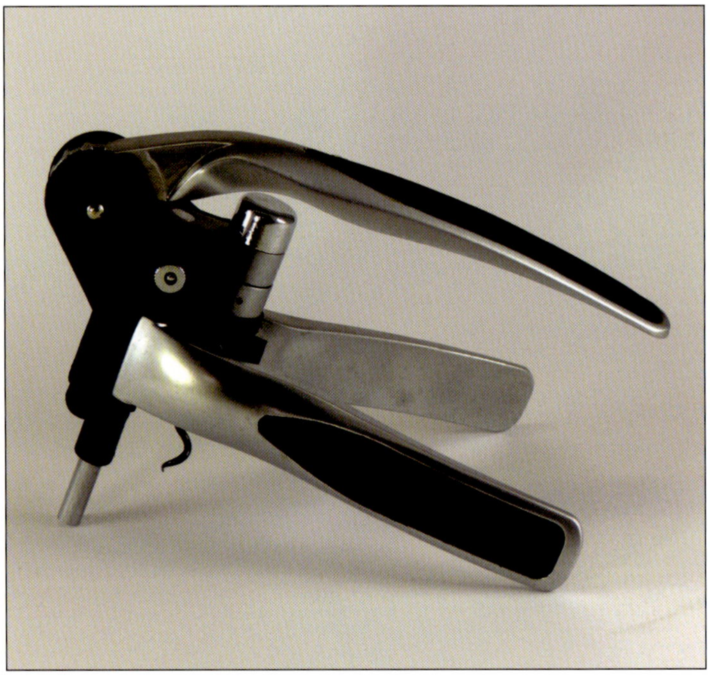

Kurketrekker, Model 106.

Kurketrekker, Model 103.

Kurketrekker, model?, (a good copy of the Elegance).

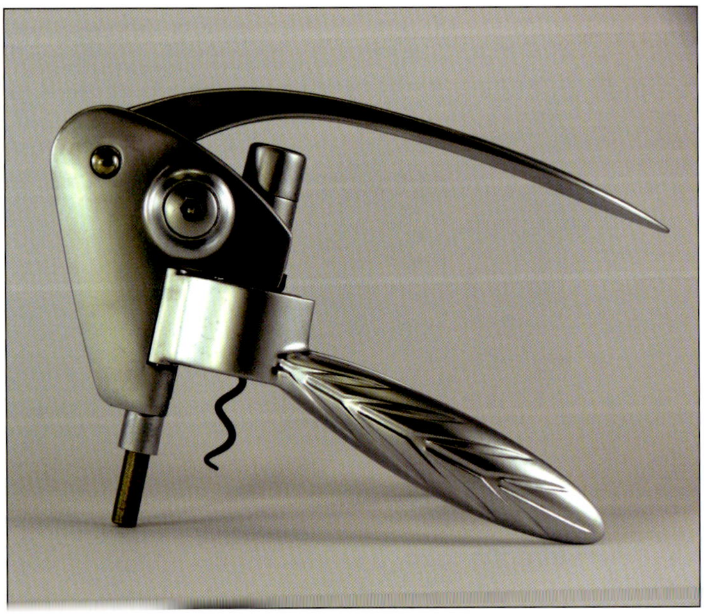

No name, heavy metal with a beautiful natural leather case.

Houdini, marketed by Metrokane, New York, New York, design by Pollen, New York, New York, made in China, 2002.

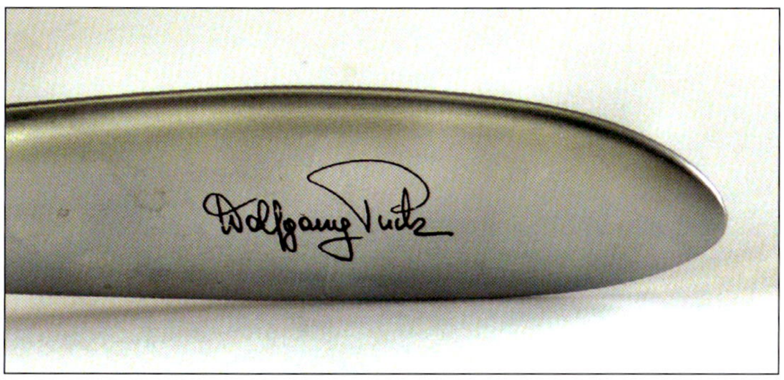

Heavy metal with "Wolfgang Puck" inscribed on top of pull lever, made in China.

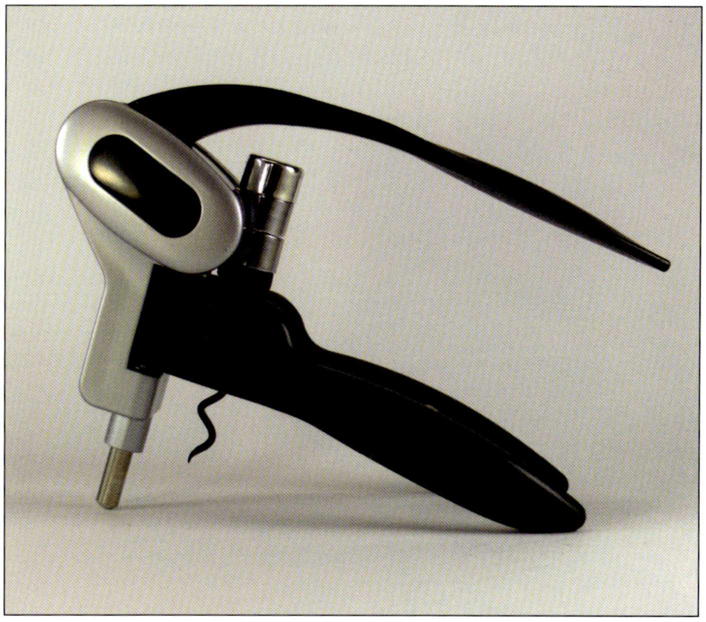

Swiftpull by Franmara.

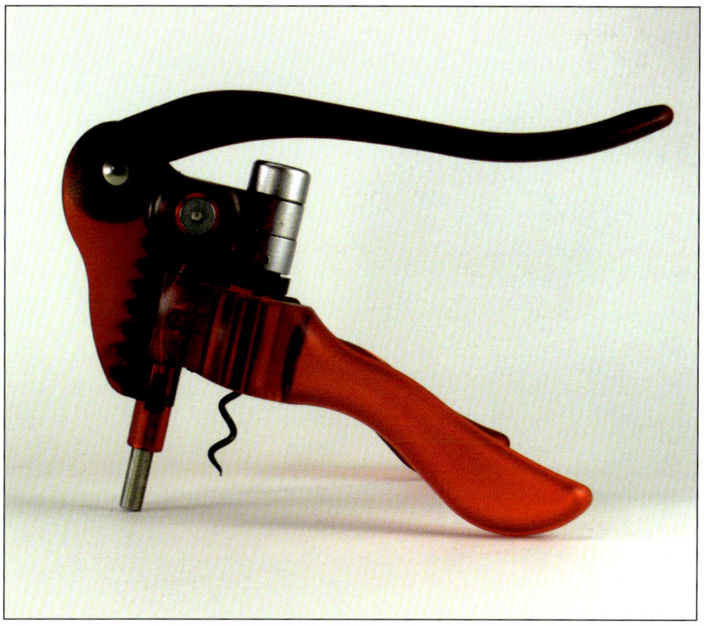

Zoom, transparent clear plastic, made in China, 2002. Other colors include blue, red, black, and silver. This version is also produced under the name of Corkscrew, iLever Model, "The Master of Life."

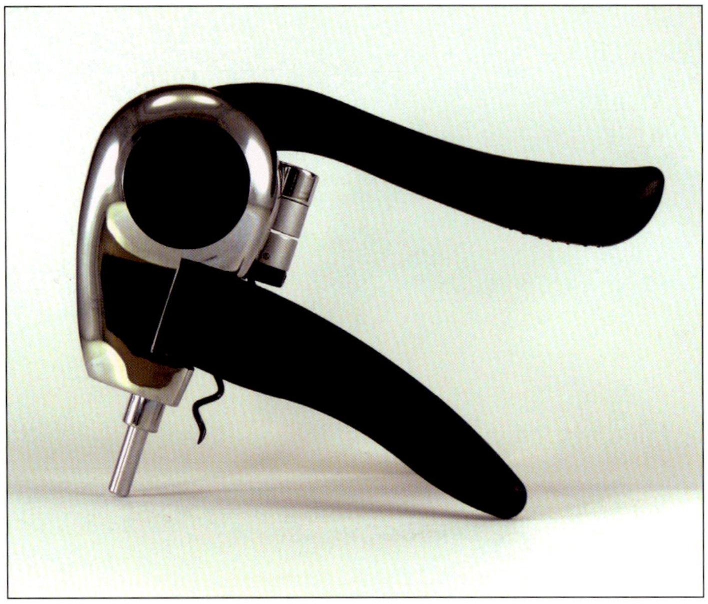

Heavy metal and plastic by Sharper Image.

Corkscrew, black plastic, made in China, 2001.

The Brookestone, black/gray plastic. The grip handles and lever are made with a honey-comb type construction.

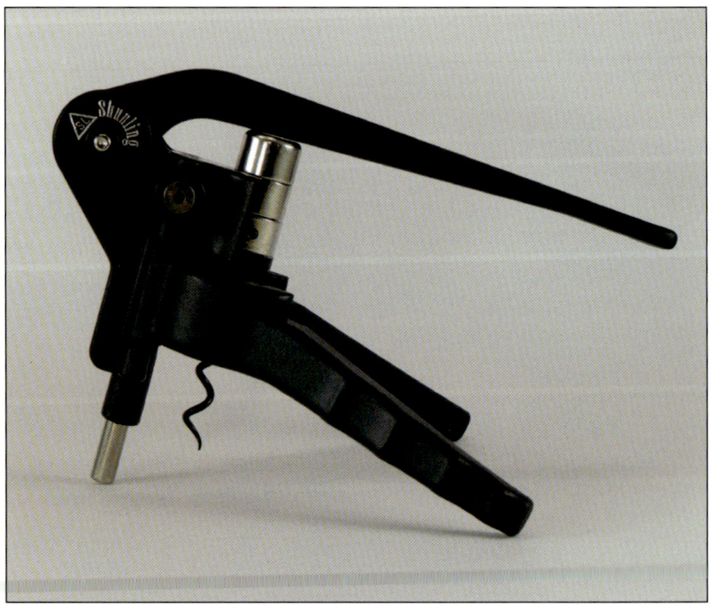

The Shunling, black plastic, made in China, 2001. An identical version of this corkscrew is sold as the Concorde (without the Shunling logo).

Unmarked vertical type lever corkscrew, well made, c. 2004. A unique idea using the foil cutter designed as a stand for the corkscrew.

The Turning Leaf lever-operated corkscrew distributed by The Turning Leaf Winery, California, c. 2001.

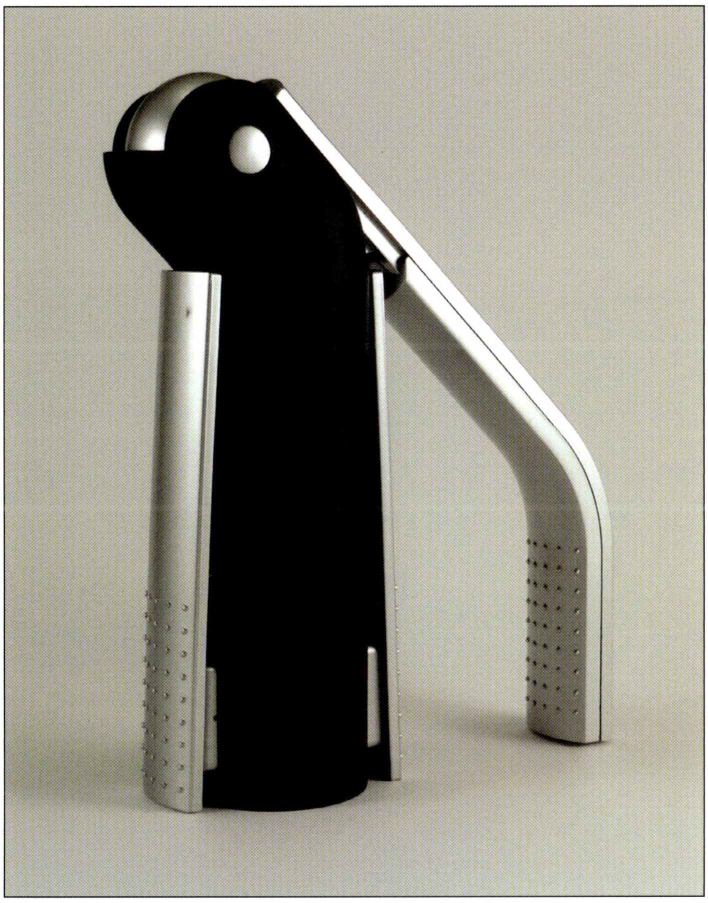

The Corklifter, also distributed as The Spacelight, vertical grip pressure insertion type corkscrews with a very similar action as the other Leverpull type corkscrews. These were made in China, c. 2005.

The Trudeau designed by Claude Mauffette, made in China, c. 2001. The foil cutter fits into a slot in the side of the stand.

The Penguin - ez Pull, made in China, c. 2005.

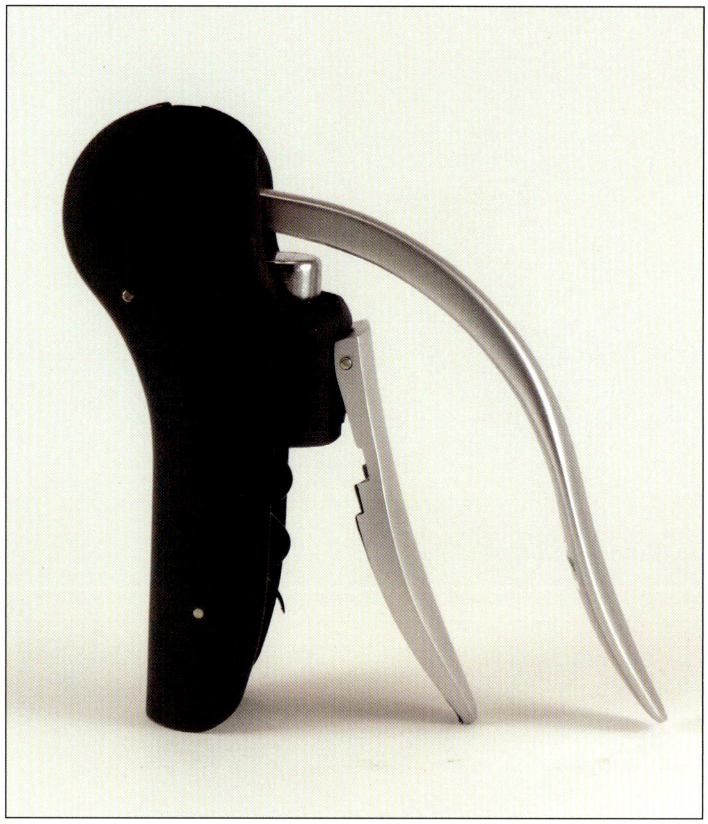

The Wine Connoisseur, Cost Plus, Inc. World Market, made in China. Also marketed as The Professional Corkscrew by Living Home. Also, one that is near identical is called Luxury Corkscrew by Tchibo, GmbH Germany. Also an unmarked heavy chrome plated metal version similar to the True Nautilus.

The Rocket, heavy chrome-plated metal, made in China.

The Orka by Mastrad, Inc., made in China, c. 2004.

The Perfect Corkscrew S.L., pro-idee, made in Madrid, Spain, c. 2001.

A chrome-plated metal version of the Wine Connoisseur, similar to the True Nautilus.

The Dolphin, made exclusively for *Wine Enthusiast*, c. 2005.

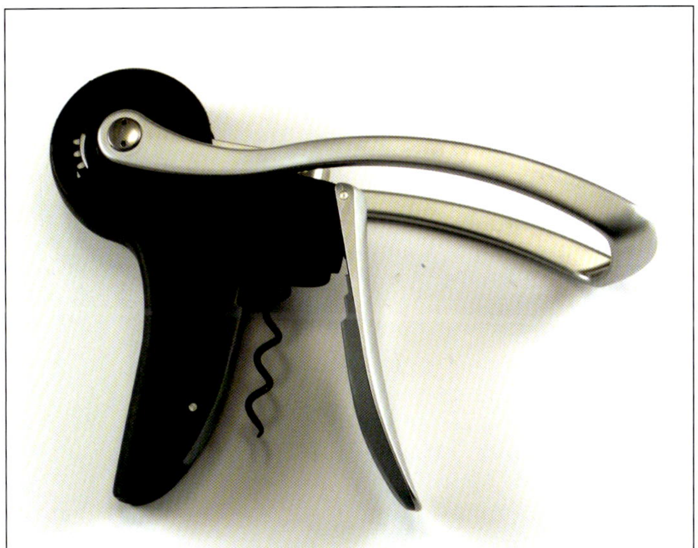

The Vigneto, made in Italy by Ghidini for Williams Sonoma, c. 2002. Also marketed as The Enoteca Uncorking Machine, distributed by Franmara, Inc.

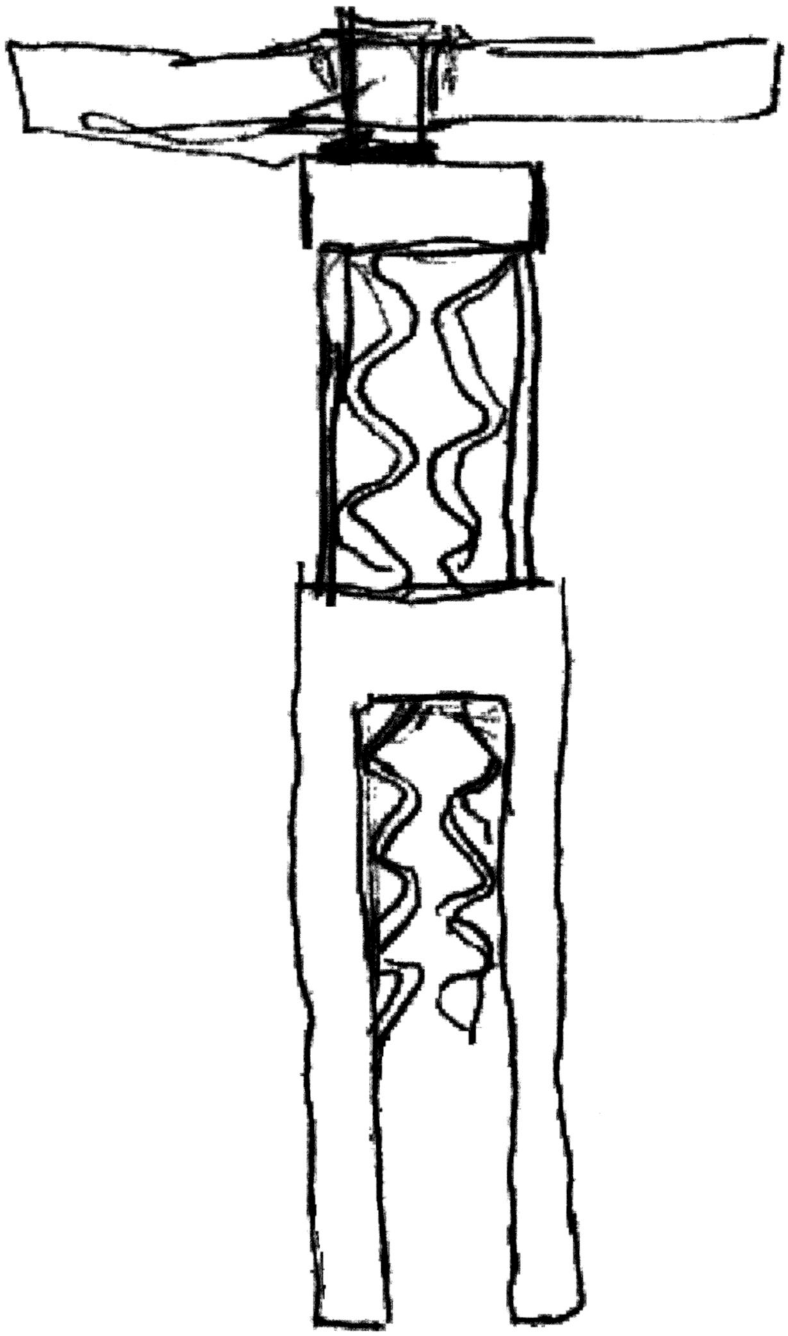

Courtesy of HA

Chapter 16: Other Related Items

A Worldwide Market

On a trip through the wine countries of Germany and France I saw a show window in a small wine shop in the small Medieval village of Riquewihr, Alsace, France, on a Sunday afternoon. A corkscrew collector can spot a window like this from a block away. The Pugipull (the jack-up corkscrew from Portugal) array around a cobalt blue bottle on the right, a Zig-Zag on the left, waiters' screws all over the place, several two-prong cork lifters, a folding bow up front, a display board with an array of Pulltek waiters' screws on the top left, and an entire board full of reproduction antique corkscrews as a backdrop. All of these and many more not in the photo, in one shop, is a collector's dream, but, for the "Screwpull Guy," the four Table Model Screwpulls in the center is the ringer. There are three TM-100's, in black, green, and blue, standing up in their cases. The case of the blue one is resting on top of the fourth one lying on a horizontal case containing a Screwpull Table Model in the burgundy color. Since the shop was closed on Sunday I was not able to go in. If you look closely, the card inside the Table Model cases is quite different from the ones produced by the Hallen Company. Being the year 2005, these were produced by LeCreuset, and appear to be Version 3 of the production models, thousands of miles from home. Unfortunately, I did not have one of these Screwpulls with the newer card. The Hallen Company model numbers were four digit numbers, ie: 1010, 1020, 1030 designating the Table Model style and the color. TM-100, a LeCreuset model number, indicates that it is a Table Model of this style or design, no matter what color.

The two ladies that ran the show at The Hallen Company discussing business over a pub lunch in London, c.1988. On the left is Vicky Kaufman, Director of Hallen Company, Ltd. UK, and on the right is Loretta Haren, Vice President of The Hallen Company, Houston. After The Hallen Company transferred ownership to LeCreuset, Vicky stayed on in London for several years to help make the manufacture of Screwpull products a smooth transition.

Marketing

Pictured below are Loretta Haren and the late George Fuermann (1918–2001) a columnist and author of "Wine Talk," a column in the *Houston Post* newspaper, from 1984 to 1995. Loretta Haren was the vice president of the Hallen Company from the beginning to the LeCreuset transfer. The photo shows them at Tony's Restaurant in Houston, where George is gathering information for an up and coming article on the Screwpull. George Fuermann was a highly respected author, in his early days, of historical facts about Texas, Houston, and the people who made it all happen. In his later years, everyone looked forward to his next "Wine Talk" column, where his love and knowledge of wine was evident.

Screw Coating Company

The Thermech Coatings Company in Anaheim, California, did all of the early DuPont "Teflon" coatings of the corkscrews for the Hallen Company, from 1979 to 1984.

An historical opening of the $31,000 bottle of Chateau Lafite Rothschild, 1822. It was purchased by John Grisanti, a restaurateur from Memphis, Tennessee, in July of 1981. The occasion was an auction benefit for St. Jude Children's Hospital in Memphis. At the opening of the bottle, Mr. Terry Robards was given the honor to wield the cork puller. He remarked that he considered only one tool for this job, the Screwpull, and it performed as expected, leaving not a crumb of cork in the bottle. Mr. Terry Robards, left in the photo, (Grisanti right) was the Wine Critic for the *New York Times* at the time. It was later reported that Mr. Grisanti was just one of ten in a partnership that purchased the wine.

"I Had to Try It"

Herb Allen said that the Teflon-coated self puller Screwpull worked so well on old and crumbly corks that maybe, if we used two screws, it would give twice the surface area needed to help pull the really, old, damp, crumbly corks without crumbling them up so badly. Just maybe.

A few weeks later, with the help of his machinist friend Roland Pair at Cameron Iron Works, Allen had a two-screw Screwpull prototype. The screws each had a gear interlocking at the top, of the same size and number of teeth, so when the handle on top of one of the screws was turned in one direction the other screw would turn at the same speed in the opposite direction.

Then came time for the test. After several dozen corks were "pulled," (they crumbled), it was decided that this device was a much better cork "grinder" than a cork puller. Allen said: "I had to try it" and then abandoned the project of two-screw Screwpulls forever.

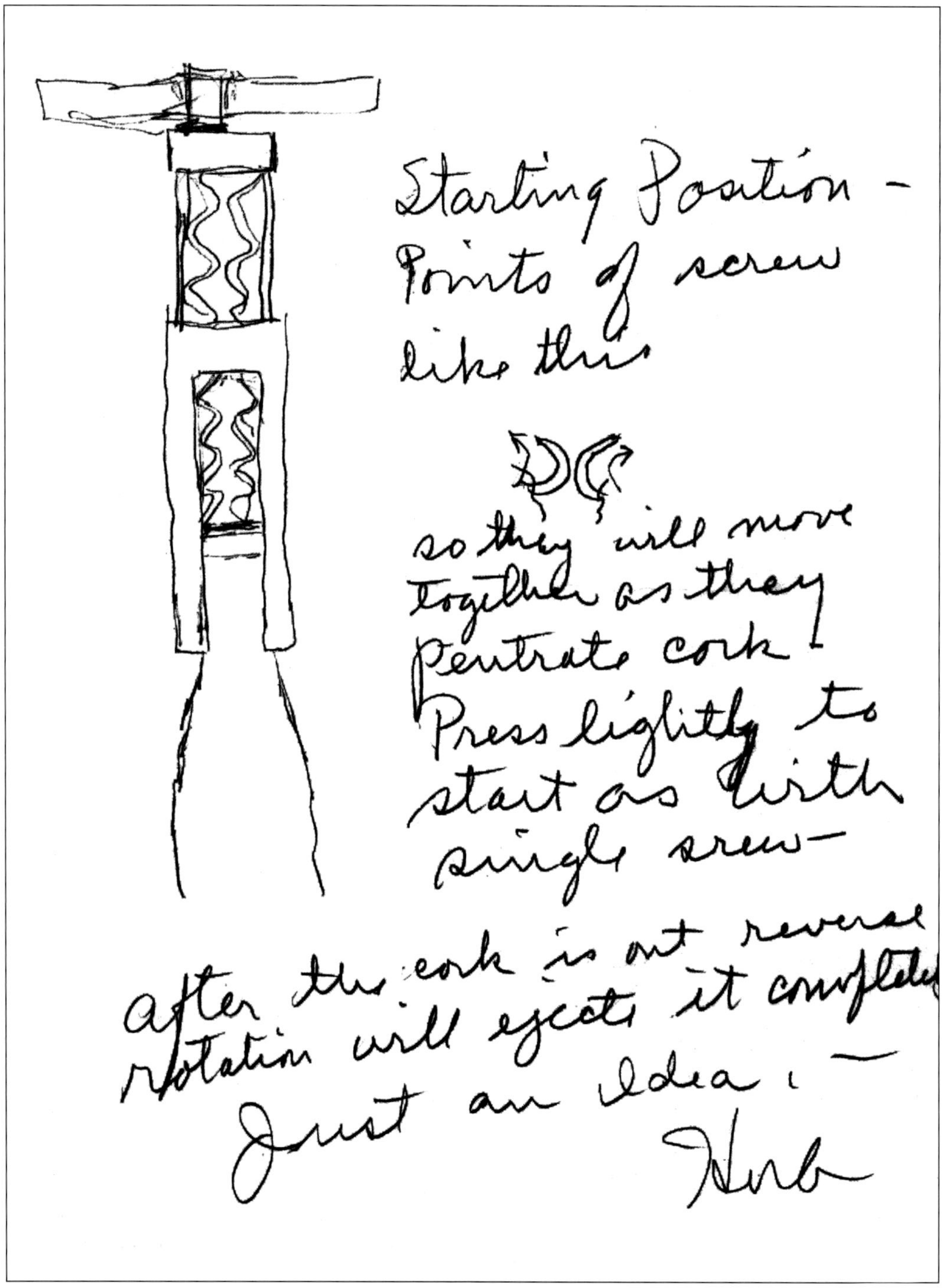

A sketch found sometime after Herb died, possibly his directions to Roland Pair.

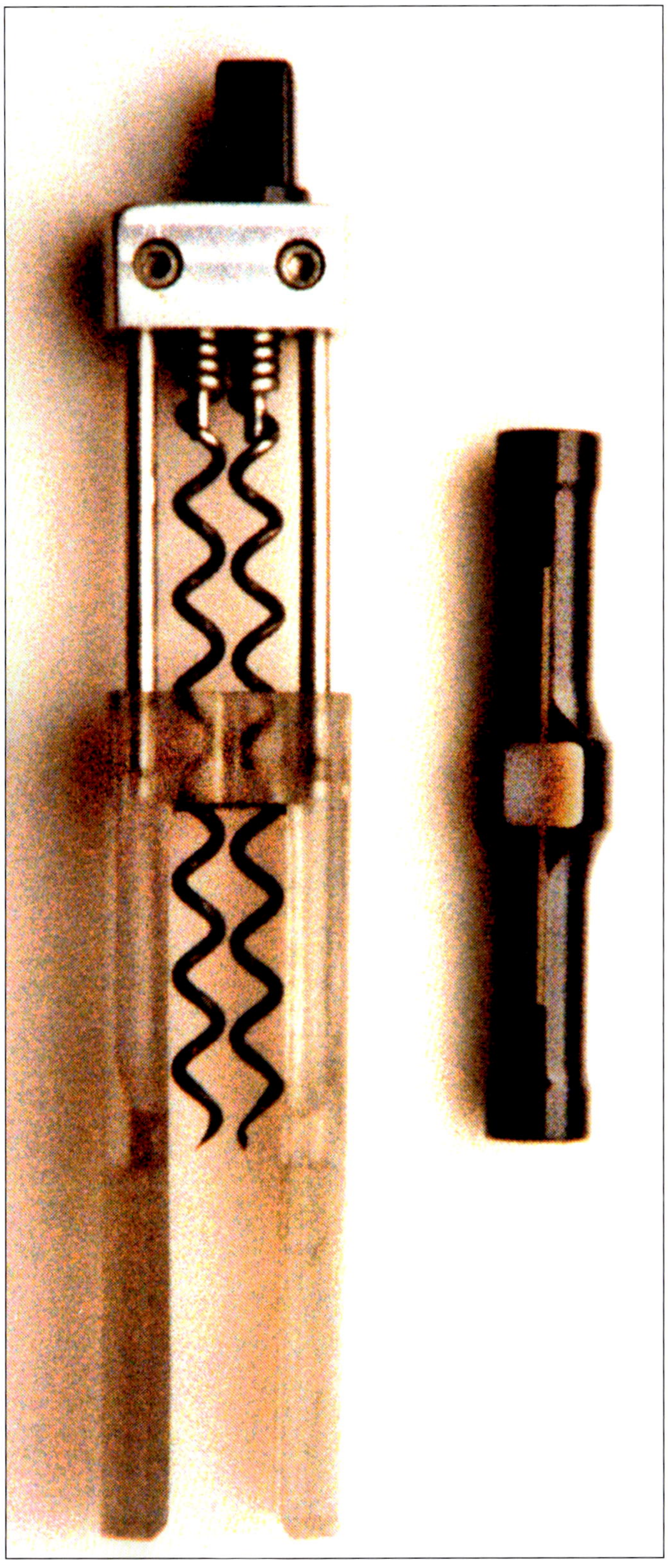

Photos courtesy of Herb Allen

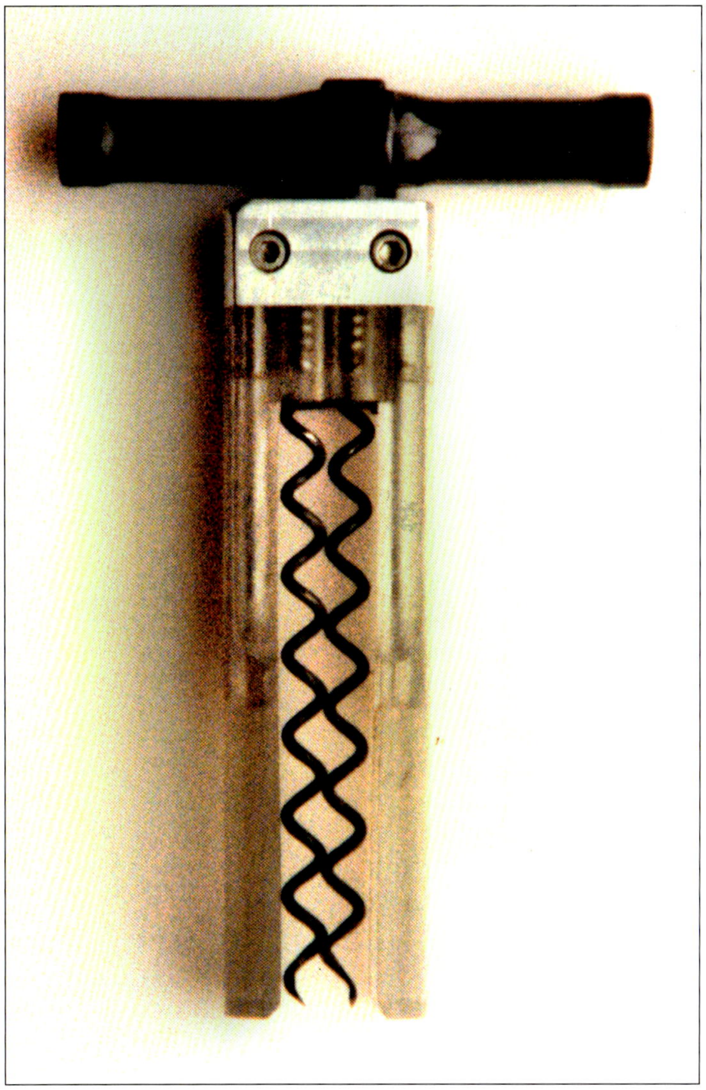

Hallen

TOOLS FOR THE GOOD LIFE®

PRICE INFORMATION

Description	Item #	Color	Unit Price/WHC*
Screwpull® Table Model	1010	Amber	$ 9.95
	1020	Burgundy	9.95
	1030	Black	9.95
	1040	White	9.95
	1090	Red	9.95
Pocket Model	2010	Black	9.95
	2020	Red	9.95
Gift Set	3000	Black	13.45
	3500	Red	13.45
	4000	White	13.45
Champagne Star (12 pack minimum)	5000	Black	5.75
	5001	Red	5.75
Foilcutter (12 pack minimum)	6000	Black	3.50
	6100	White	3.50
	6200	Black (Blister Pack)	3.50
	6300	Red	3.50
Lever Model & Foilcutter Set	7000	Black	70.00
	7001	Replacement Screw	8.00

**Prices subject to change without notice. Patented and Patents Pending*

TERMS AND CONDITIONS

To Order: Call toll free 800-231-2746, or (713) 683-8111.
Send purchase orders to: Hallen Company, P.O. Box 1392, Houston, Texas 77251-1392.
Minimum order $180.00.

Custom Imprinting: $125.00 tool and die charge; $0.25 per color, per unit;
$50.00 re-order charge.

Delivery: Shipments under 100 pounds will be shipped FOB, Houston, Texas via UPS unless specified otherwise.

Payment Methods: New account—COD, FOB, factory, Houston, Texas.
Add $3.30 for each COD package.
Established account—Net 30 days with credit approval.

Returns: Authorization number required. A 20% handling and restocking fee will be charged. Returns must be sent freight pre-paid and be received in good condition. Custom imprinted goods cannot be returned.

Shipping And Handling: Orders up to $200–$5.00; $201-$350–$8.00; $351-$500–$10.00; $501-$1,000–$15.00.
Orders over $1,000, please contact Hallen Company. Hawaii, Puerto Rico, Virgin Islands, and Caribbean are excluded from these charges.
C.O.D.: $3.30 per package, in addition to the above rates.
For shipments other than UPS, including overnight (UPS blue and red), please contact Hallen Company.

	Description	Dimensions As Packaged	Weight
Individual Product:	Screwpull® Table Model	6½" × 2¾" × 2"	5 ounces
	Pocket Model	6" × 1" × 2"	4 ounces
	Foilcutter	2" × ½" × 2⅞"	1 ounce
	Lever Model & Foilcutter Set	6½" × 9" × 3"	17 ounces
	Gift Set	6½" × 2¾" × 2"	6 ounces
	Champagne Star	2" × 1⅞" × 1⅞"	4 ounces
	Foilcutter Blister Pack	7⅝" × 4⅜" × ⅜"	1 ounce

Hallen Company—Houston
P.O. Box 1392 Houston, Texas 77251-1392
Telephone:
800-231-2746 or
(713) 683-8111
FAX: (713) 683-9546

Hallen Co., Ltd.—London
Southgate, London NII 1JL, ENGLAND
Telephone:
44-81-368-8318
FAX: 44-81-361-4480

"Tools for the Good Life" and "Screwpull" are registered trademarks of the Hallen Company.

Hallen

TOOLS FOR THE GOOD LIFE®

Wholesale price list, c. 1980.

Screwpull screws in racks at the manufacturing company.

PLAYBOY

ON·THE·SCENE

WHAT'S HAPPENING, WHERE IT'S HAPPENING AND WHO'S MAKING IT HAPPEN

BALI'S SILVER HIGH

When silver designer John Hardy set out 20 years ago on a trip around the world, he never got past Bali. The island's beauty and traditions inspired him to settle and to begin training local artisans to craft raw silver into masculine jewelry and accessories. Today, the John Hardy Collection numbers more than 1000 different objects, ranging from a money clip and a gentleman's flask to a cigar ashtray and a corkscrew (all shown below). (You may remember Hardy's cigar tube and lighter from last December's *Christmas Gift Guide*.) Created almost entirely by one talented artist, each piece in the collection is masculine yet delicate, with its burnished silver patina and a variety of intricate details. The result? A powerful look and feel.

The John Hardy Collection of Balinese silver jewelry and accessories includes his Kuno-Jawan–pattern cigar ashtray, which is 6½" in diameter and includes a stogie rest ($610). At far right is a small flask that's perfect for your jacket and holds about four ounces of your favorite liquor ($495). The wine pull ($685), which comes with a palmwood base (not shown) and handle, is positioned next to Hardy's Pintusapi-pattern money clip ($190). At center: A silver carved-chain ID bracelet ($415) encircles some premium double corona Playboy Cigars by Don Diego.

An article in the April 1997 issue of *Playboy Magazine* by James Imbrogno about the John Hardy Screwpull.

Screwpull®

The registered Screwpull logo.

The big three of Richard's Liquor Stores, Houston, Texas, c. 1979. Left to Right: cellarmaster Henryk Kucharzyk, owner Richard Trabulsi, winemaster Bob Roy.

According to Herb Allen's secretary, Virginia, Herb had 24 gold Screwpull miniatures made and presented to people that he considered to be helpful and/or instrumental in helping him make his "idea" of the Screwpull family become a reality. They were made by a jeweler in New York City. The miniatures stand a full 1 1/4" tall on their stands. Shortly after Herb was inducted into the ICCA, as an Honorary Member, he called and wanted to borrow my ICCA badge for a short while without letting me know why, so I sent it over without question. In a matter of a few days the badge was returned and a few days later a carton of four Screwpulls arrived in the mail with our ICCA logo imprinted on them. He sent each of our 50 members four of the Table Models of different colors with the logo on them. He considered the title and honorary membership in the ICCA to be such a great honor that he sent the Screwpulls out as a show of appreciation. Nearly 30 years later, they are still a very treasured gift to the surviving members.

A solid gold, miniature, operational Screwpull.

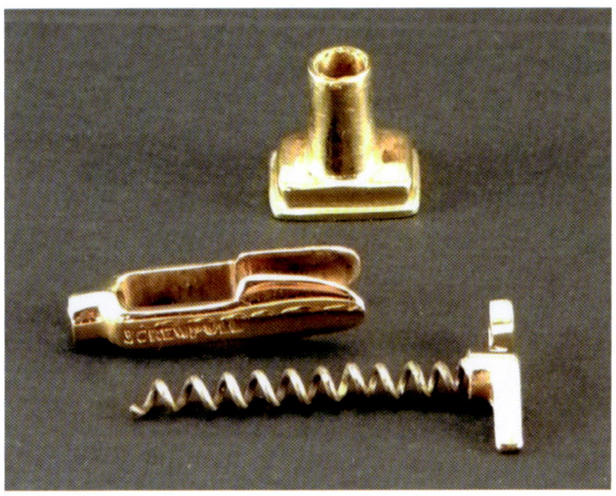

The Last Word

Only time will tell how many new reproduction designs will come out of Herb Allen's Leverpull. If the trend of the wineries is to use more and more dense composite and plastic corks then the end is near for the direct pressure Leverpull type corkscrew. If artificial corks are developed that can be penetrated by lever type corkscrews, then who knows how long they will be around.

Seeing how much cork is being produced in Portugal and Spain, I really do not see a shortage of real cork in the near future for the sealing of wine bottles. Apparently, the reasons the wineries are drifting away from real cork is cost and that some corks will taint and spoil a percentage of the wine. Both are valid reasons in the wine industry today, so a major change is inevitable. Depending on how dense and hard the replacement seals become will determine the fate of the Lever style corkscrews. Then, if more and more screw caps become the norm instead of cork type wine bottle seals (the wine industry of New Zealand is already at 90%+ screw caps), the fate of all corkscrews is in jeopardy.

In front of all your guests at a very special festive dinner, out comes a pair of channel-lock pliers from the kitchen drawer to help arthritic fingers coax a stubborn screw cap from a great bottle of wine—how romantic is that? I found myself in that very position, without a pair of pliers, and discovered that a Screwpull Foil Cutter will cleanly take the top right off of an aluminum screw cap or twist off cap. I will be willing to bet that Herb Allen did not know that he had already solved that problem.

And finally, I am sad to hear that 2012 will be the last of the Screwpull logo in the US and Canada as well as many other countries. Screwpull has been 33 years on the world market and from now on the products will be known as Le Creuset. That comes as quite a shock to the "Screwpull Guy," but life goes on.

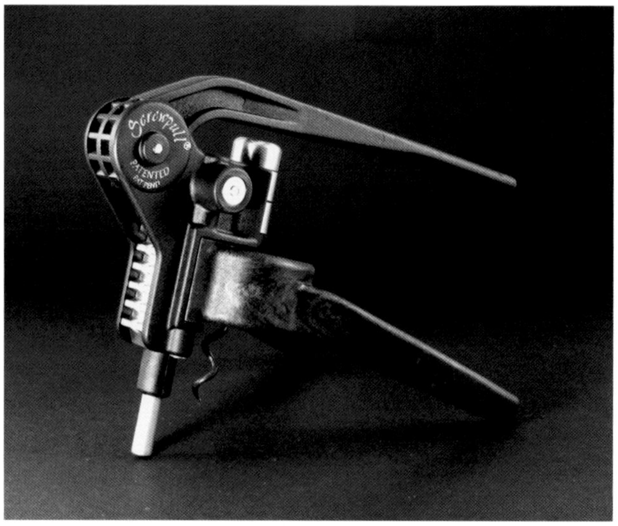

Screwpull World Record: The Screwpull holds the Guinness Book of World Records title for most wine bottles opened in one minute—including foil removal.

Other Schiffer Books by the Author:

Ring Bit: History, Form, & Function, 978-0-7643-3887-8, $49.99

Other Schiffer Books on Related Subjects:

The Ultimate Corkscrew Book, 0-7643-0701-0, $89.95

Wine Antiques and Collectibles, 978-0-7643-4335-3, $79.99

Champagne Collectibles, 978-0-7643-3721-5, $79.99

Library of Congress Control Number: 2013935649

Cover Design: John Cheek
Type set in Minion Pro

ISBN: 978-0-7643-4421-3
Printed in China

Published by Schiffer Publishing, Ltd.
4880 Lower Valley Road
Atglen, PA 19310
Phone: (610) 593-1777; Fax: (610) 593-2002
E-mail: Info@schifferbooks.com

For our complete selection of fine books on this and related subjects, please visit our website at www.schifferbooks.com. You may also write for a free catalog.

This book may be purchased from the publisher. Please try your bookstore first.

We are always looking for people to write books on new and related subjects. If you have an idea for a book, please contact us at proposals@schifferbooks.com

Schiffer Publishing's titles are available at special discounts for bulk purchases for sales promotions or premiums. Special editions, including personalized covers, corporate imprints, and excerpts can be created in large quantities for special needs. For more information, contact the publisher.

SCREWPULL